North American Mixing Forum

AA002662

Held at the 2024 AIChE Annual Meeting

San Diego, California, USA
27-31 October 2024

ISBN: 979-8-3313-1664-8

North American Mixing Forum

Held at the 2024 AIChE Annual Meeting

San Diego, California, USA
27-31 October 2024

AIChE
The Global Home of Chemical Engineers

Printed from e-media with permission by:

Curran Associates, Inc.
57 Morehouse Lane
Red Hook, NY 12571

CURRAN ASSOCIATES INC.
proceedings
.com

Some format issues inherent in the e-media version may also appear in this print version.

Printed with permission by Curran Associates, Inc. (2025)

For permission requests, please contact AIChE
at the address below.

AIChE
120 Wall Street, FL 23
New York, NY 10005-4020

Phone: (800) 242-4363
Fax: (203) 775-5177

www.aiche.org

Additional copies of this publication are available from:

Curran Associates, Inc.
57 Morehouse Lane
Red Hook, NY 12571 USA
Phone: 845-758-0400
Fax: 845-758-2633
Email: curran@proceedings.com
Web: www.proceedings.com

TABLE OF CONTENTS

COMPUTATIONAL ANALYSES OF MIXING PROCESSES I

COMPUTATIONAL ANALYSES OF MIXING PROCESSES II

COMPUTATIONAL ANALYSES OF MIXING PROCESSES III

EXPERIMENTAL INVESTIGATIONS OF MIXING PROCESSES

MIXING IN BIOTECHNOLOGY, FOOD AND BEVERAGE, AND AGRICULTURE SECTORS

MIXING ISSUES IN SCALE-UP AND SCALE-DOWN OF PHARMACEUTICAL AND BIOPHARMACEUTICAL PROCESSES

MIXING IN MULTIPHASE AND REACTIVE PROCESSES

DIVISION PLENARY: NORTH AMERICAN MIXING FORUM AWARDS SESSION (INVITED TALKS)

417e Convective Heat Transfer in Turbulent Flows

John Thomas, Eric E. Janz, Brian DeVincentis

A generalized approach for predicting turbulent convective heat transfer coefficients using large eddy simulation is presented. This model is based on semi-empirical transport theory, which links the local convective heat transfer rate to the local energy dissipation rate within the near-wall boundary layer. We apply these approaches to agitated tanks over a range of operating conditions and impeller types, as well as pipe flow systems, cylinders in crossflow, and tube bundles. We validate the generality and reliability of the theoretical models against expectations from experimentally derived empirical design correlations.

The findings from this investigation present thee key implications: first, as a complement to physical testing, this modeling approach can be used to reliably predict local and overall heat transfer rates in agitated/turbulent systems using first-principles simulations. Second, because the approach appeals directly to the Navier-Stokes equations, it provides mechanistic insight into the link between operating conditions, fluid flow, and thermal convection. Third, because heat transfer is modeled in tandem and on-the-fly with the underlying fluid dynamics, the combined effects of convection and advection and be investigated directly, in terms of temperature dependent viscosities and reaction rates. In this sense, the approach can be used to perform first principles multi-physics simulation that make a priori predictions of convective heat transfer rates and thermal transport in turbulent flows.

A general approach for predicting convective heat transfer coefficients in turbulent systems, International Journal of Heat and Mass Transfer 220(24):124989

83h Non-Newtonian mixing; select case studies from lower consistency to ultra-high

Hossam Metwally, Sravan Nallamothu, Prashanth Ethirajulu

In this presentation, three separate case studies will be presented.

In the first case study, the flow of a power-law non-Newtonian liquid with relatively low consistency in a flat bottom baffled tank is simulated. Emphasis will be on predicting axial, radial, and tangentially velocity distribution and comparing it with available experimental data.

The second case study will have a higher consistency index. The vessel will be equipped with double helical blades. Focus will be on predicting blend time, torque/power, and the viscosity field.

Finally, the third case study will have a gel-like, highly viscous non-Newtonian, power-law, liquid flowing through an eight-element static mixer. The case will focus on the viscous heating aspect associated with both the high viscosity and the static mixer geometry.

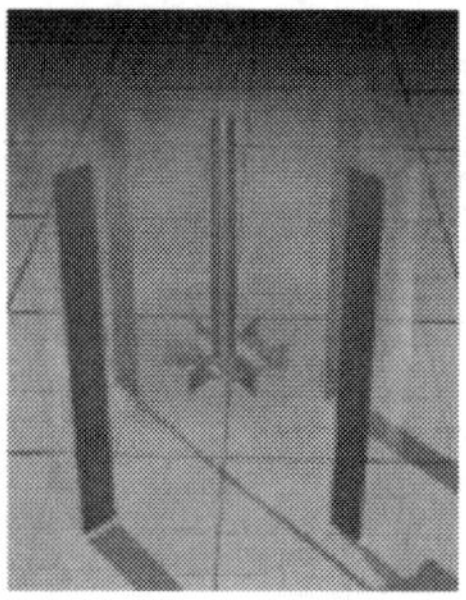

Case Study I

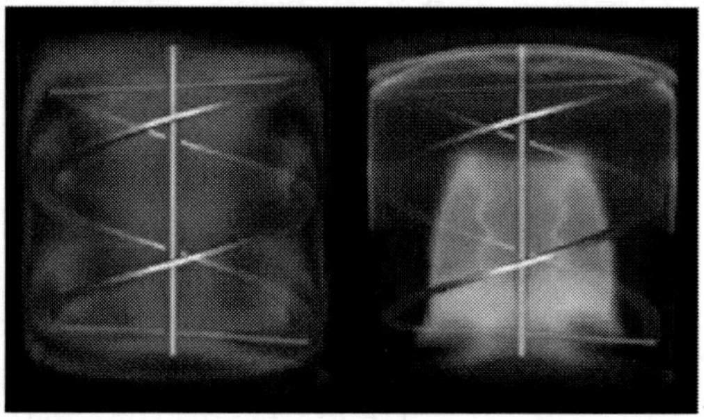

Case Study II

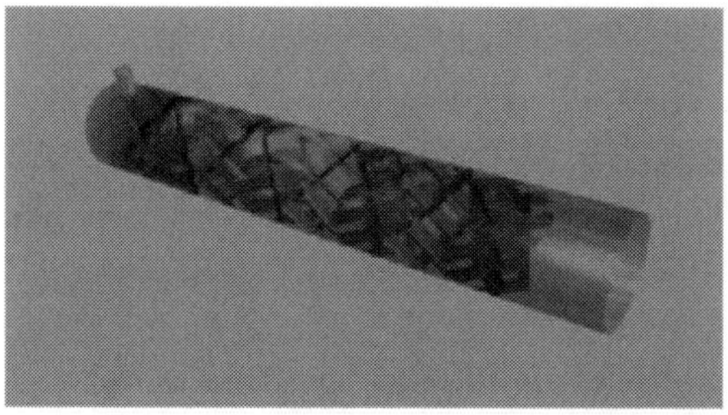

Case Study III

83g Apparent Suspension Viscosities Found By Computationally Simulating a Couette Viscometer

Sudarshan Srinivasan, Harry Van den Akker, Orest Shardt

We report the results from direct numerical simulations of the flow of slurries in a Couette viscometer. We used a lattice Boltzmann technique with an immersed boundary model for studying the fully resolved response of a slurry to a simple shear as in a Couette viscometer. The slurries consisted of a Newtonian liquid carrying mono-sized spherical solid particles in the range 150-300 micron, up to solids volumes as high as 50%. We used either short-range spring forces or lubrication forces to deal with particles approaching each other closer than the lattice (grid) spacing. We computed the relative apparent viscosity for several shear rates and particle concentrations, and discuss the effects of these variables on particle rotation and cluster formations. The apparent viscosities increase with increasing particle Reynolds number (shear thickening) and solids fraction. As long as the particle Reynolds number is low (0.1), the computed viscosities are in good agreement with experimental measurements as well as with theoretical and empirical equations. To mimic the flow behaviour of iron oxide particles in a caustic medium, we added electric double layer forces to the simulation to model the pertinent repulsion between the particles. We report on the effect of pH in the range 9 through 12 and of the Debye length on spatial particle structures, resulting in a reduced apparent viscosity at higher solids volume fractions.

83d Numerical Investigation of Mixing Strategies in Multiphase Loop Reactors

Malik Hassanaly, Pranav Jain, Mohammad Rahimi, Devon Sigler, Hari Sitaraman

Sustainable fuel production is a promising strategy to help mitigate the carbon footprint of transportation sectors such as aviation that are difficult to electrify. In particular, gas fermentation of CO2 along with renewable hydrogen and syngas, is currently being explored as a viable pathway for the production of sustainable aviation fuel intermediates, from captured CO2.

The performance of gas fermentation reactors at-scale (~ million liters) strongly depends on the efficient mass transfer of species from the gas to the liquid phase. To address this challenge, we investigate loop reactor designs with strategically placed static and dynamic mixing elements for improving mass transfer at scale. Loop reactors (Fig. 1) typically are more compact than stir tanks, and enable greater control of gas residence times. However, the design of loop reactors at scale, that simultaneously achieves adequate mass transfer of gas mixtures and reduces inhomogeneous species distribution is a challenging research problem.

In loop reactors, mixing is best achieved through the strategical placement of static/dynamic mixers and spargers. The objective of this work is to optimize the spatial locations of mixers and spargers to maximize interphase mass transfer while minimizing power consumption. To solve this optimization problem, we will use an ensemble of simulations of loop reactor designs using our validated high-fidelity multiphase computational fluid dynamics (CFD) model (Rahimi et al., Chem. Eng. Res. Design, 139, 2018, https://github.com/NREL/BioReactorDesign). The effect of the placement and the type of mixer adopted will be described, and an optimal mixing strategy will be described.

Figure 1: Liquid velocity distribution in a loop reactor similar to (Lackner et al., Natural Gas—New Perspectives and Future Developments., 2022)

83e Performance Improvement of in-Line Flow Chemical Reactors By Modifying Injector Geometry

Smita Agrawal, Kumar Aanjaneya

Continuous-flow or in-line reactors/mixers are ubiquitous in food, beverage, chemical, pharmaceutical, petrochemical, biotechnology, and other industries. They are used to mix two separate fluid streams with the aim to either get a homogeneous mixture or attain a chemical reaction. One of these streams is generally brought in through a "Stinger" or an additive injector which is oriented perpendicular to the direction of the main flow in the reactor and has a circular cross section. This leads to a flow structure which is similar to the canonical case of crossflow around cylinders with flow separation and vortex formation downstream. This work utilizes a combination of computational and experimental tools to study the aforementioned flow around the injector and its influence on the residence time of the additive. It is found that the additive gets "trapped" in a recirculation zone if its injection velocity is low and the injector holes are aligned parallel to the axis of the main reactor pipe. This problem can be solved by (i) increasing the injection velocity, or (ii) orienting the injector holes at an angle to the main pipe axis; such that the additive is injected away from the downstream vortices. Some results are also presented for a novel injector design with a streamlined cross section to delay the onset of flow separation. These results are important with regards to setups where unintended long residence times can cause promotion of undesirable secondary reactions. "Fouling" or deposition of gels/solids is a common symptom of such reacting flows with issues in mesoscale mixing.

83f CFD Modeling of Impeller Effects on Heating Times in a Dimple-Jacketed Stirred Tank

Richard Cope, Kevin Logsdon, Richard Kehn, Eric E. Janz

Process media in an agitated vessel are commonly heated or cooled through heat exchange with a utility fluid flowing through the vessel jacket. Agitation homogenizes media composition and temperature, and also enhances overall heat transfer through the vessel wall. Existing correlations (e.g., see J.R. Nunhez, Chapter 14, Advances in Industrial Mixing, pp 491-532, John Wiley & Sons, New York, NY, 2016) help estimate heat transfer coefficients (process fluid, utility fluid, and overall) and heating / cooling times. These correlations are based on fluid physical properties, and also fluid behavior as produced by given mixing impeller types.

This work compares laboratory results with M-Star CFD (computational fluid dynamics) predictions. Thomas et al. (Thomas, J. A., DeVincentis, B., Janz, E., & Turner, B., 2024, Int. J. Heat Mass Transf, Vol. 220, p. 124989, Elsevier) introduced the M-Star CFD generalized method of estimating convective heat transfer coefficients using local fluid properties (e.g., viscosity) and operating conditions (e.g., energy dissipation rate) near the vessel wall. While they validated the method against various experimentally derived empirical design correlations, this comparison with experimental data provides greater understanding of the method's generality, validity and capability.

M-Star CFD simulations and laboratory results are compared while assessing the impact of three different impellers (i.e., narrow hydrofoil (A310), Rushton turbine (R100), or 4-45 PBT (A200)) on heating times for 21gals water in a dimple-jacketed 18.1-inch stainless steel vessel. During each laboratory test, oil flow rate, time, and temperatures (water and oil) were recorded each second. The lattice Boltzmann LES solver in M-Star CFD provides transient flow and thermal field results while modeling process and jacket fluids, and temperature of the separating wall. This enabling comparison of model and measurement, as a function of time, facilitates closer interrogation using predicted and measured temperature profiles.

145a Coalescence Re-Dispersion Model of Spielman and Levenspiel and Interaction By Exchange with the Mean Model of Villermaux Revisited

Damaraju Phaneswararao, Dhiraj Garg, Pedamalli Reddy

The CD model of Spielman and Levenspiel and interaction by exchange with the mean(IEM) model for prediction of micro-mixing effects in a CSTR are reviewed. An interpretation of the interaction parameter of the CD model is given by using the random walk theory of Chandrasekhar and the results obtained by Rao and Dunn on Monte Carlo simulation of coalescence re-dispersion phenomena to simulate mixing with homogeneous reactions in a plug flow reactor with unmixed feed. The concentration changes simulated by the CD model are proved to be identical to the concentration changes predicted by the differential equation for unsteady steady state diffusion based on Fick's law when written in finite difference form. The application of the CD concept to the solution of the film model for diffusion with reaction in a gas-liquid film also is demonstrated. The extension of the plug flow reactor model of Rao and Dunn to predict recycle flow effect on RTD of a CSTR shows similarity in predictions of RTD data with those observed by Leitman and Ziegler. Aubry and Villermaux claimed that the results obtained by using IEM model are nearly identical to those results obtained by using the CD model if the equivalence relationship between the parameters of the models proposed by them is used. In the present work, it is observed that the effect of volumetric feed ratio of reactants fed in unmixed condition to a CSTR predicted by the two models are widely different after the equivalence relationship proposed by them is used. In addition the effect predicted by the IEM model is found to be incorrect. A set of typical results obtained in the case of a (1,1) order irreversible reaction taking place in a CSTR with unmixed feed are given in Figure 1.

Comparison of IEM with MCM model predictions of 1-x vs $V_r/(V_r + 1)$ for (1,1) order reaction in a CSTR with unmixed feed , s=1, I=1 (b=0.5) for different values of Da

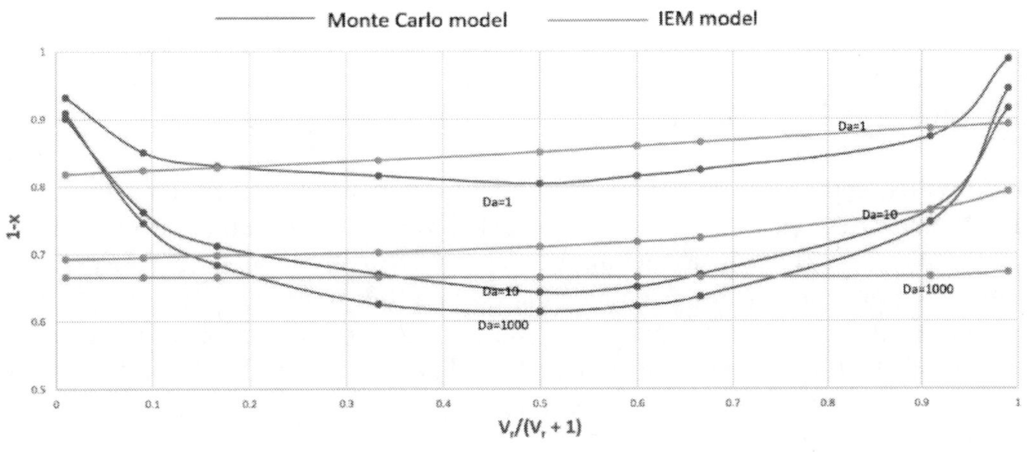

145b Computational Investigation on the Hydrodynamics of Structured Fluids in Unbaffled Stirred Tanks

Gianluca Boccardo, Antonio Buffo, Francesco De Roma, Nicola Antonio Di Spirito, Nino Grizzuti, Francesco Maluta, Daniele Marchisio, Giuseppina Montante, Alessandro Paglianti, Rossana Pasquino

Structured fluids are characterized by their capability to self-assemble and form different structures at a scale larger than the atomistic one. Many fluids fall into the category of structured fluids, like surfactant solutions, suspensions, emulsions, and solutions of polymers in water. The temperature and the composition determine the type of microstructure, which influences the macroscopic properties of the fluid, such as the rheology. For example, these fluids can exhibit Newtonian or non-Newtonian behavior depending on the microstructural phase, with abrupt variations of the local apparent viscosity. Consequently, the fluid dynamics in systems containing this kind of fluids can be strongly affected by the operating conditions, from temperature to local composition and applied strain rate. Poloxamers are nonionic triblock copolymers formed by a central chain of polypropylene oxide in between two chains of polyethylene oxide. These molecules, also known by the commercial name of Pluronics, are found in different physical states, depending on the molecular weight and the ratio between the two repeating units. The different solubility in water of the blocks constituting these polymers leads to an amphiphilic behavior, which results in the formation of microstructures in aqueous solutions. Consequently, poloxamers have surfactant properties and are widely employed in the production of pharmaceutical preparations and creams or detergents for personal care. Even though structured fluids like these are widespread, the design of the industrial processes that use them is made difficult by their complex rheology.

Nonetheless, many industrial sectors, such as food, cosmetics, and pharmaceutical ones, employ structured fluids in the production processes. These processes often involve a mixing step, in equipment in which composition and applied strain rate present substantial local variations. Stirred tanks are one of the most studied mixing equipment, with a historical focus on baffled ones, which usually assure more efficient mixing. However, a renewed industrial interest in unbaffled vessels has been growing in recent times. While the baffles help to redistribute the flux in the radial and axial direction, their absence can be a desired feature for some applications. In particular, it can be an interesting choice for fluids with high viscosity or in general for low Reynolds numbers, to reduce the formation of dead zones. Moreover, unbaffled stirred tanks are preferred for shear-sensitive fluids, as in bioreactors, and during crystallization processes to depress aggregation.

When dealing with complex models, such as the rheology of structured fluids, simulation tools can be a great aid to the design of processes and equipment, given the continuous improvements in the capabilities of software and hardware. Moreover, computational models are a way to investigate the role of the involved phenomena and to reduce the number of traditional experiments needed. This work is inserted in such context and focuses on the impact of the rheological properties of structured fluids in the previously mentioned unbaffled vessels. The aim is to help develop a clearer picture of the hydrodynamics in these kinds of equipment and to facilitate the design of future mixing processes that involve structured fluids.

Given the extensive use of poloxamers for various industrial products, the fluid chosen for this study is a mixture of Pluronic L64 at different concentrations in water. Different microstructures arise depending on the composition of this blend, as it has been already found in a previous work by Pasquino et al. [1] using experimental rheometry and dissipative particle dynamics. In general, the mixture can be considered as highly viscous fluid as the concentration of L64 goes beyond 50% in weight, which is also the limit above which a non-Newtonian behavior is observed. The choice of studying the hydrodynamics of this fluid in an unbaffled stirred tank derives from this high viscosity, together with the shear sensitivity for some applications of this product. The tank has an elliptical bottom and is equipped with a standard Rushton turbine, with a total volume occupied by the fluid equal to 10 L.

Four different blends are analyzed: pure L64 at 25°C, mixtures of L64 in water at 40% and 50% in weight at 25°C, and a mixture of L64 in water at 50% in weight at 30°C. Starting from rheological experiments, a Newtonian model is chosen for the mixtures of L64 and water at 25°C, while a power law is used for the pure L64 and the mixture at 30°C. As previously stated, the microstructures formed in such fluids are influenced also by the temperature, and from this difference in the rheology at fixed concentrations, it is possible to identify a phase transition between 25 and 30°C. The described rheology models are used in Computational Fluid Dynamics (CFD) simulations, to obtain information about the power consumption and the variation of velocity and strain rate in the system. The operating conditions in industrial equipment usually lead to a laminar fluid dynamic regimen, due to the high viscosity of these fluids. Moreover, every simulation is performed at the steady state and with constant composition of the fluid, using the multiple reference frame for the zone around the impeller. Consequently, in investigating the dependence of the power number with respect to the Reynolds number, most of the simulations performed in this work focus on Re lower than 100. The simulations conducted with the mixture of L64 40 wt.% in water, which has the lowest viscosity among the studied blends, explore higher Reynolds numbers, between the laminar and the early transitional regimen. A grid independence study is carried out for different Reynolds numbers, monitoring the power number calculated both using the torque on the impeller and the laminar dissipation energy rate.

In addition to the constitutive equations for viscosity obtained through interpolation of the experimental data with an a priori model, such as the power law, an alternative viscosity model was tested. Such a model is based on the Gaussian Process Regression (GPR), which is a statistical technique akin to machine learning. The GPR actuates a regression on the data, providing a model which outputs single values for the predicted variable, rather than a functional form for viscosity with respect to the strain rate. The GPR viscosity model takes as input the strain rate for every cell of the computational domain and returns the corresponding viscosity prediction, based on the training performed with the experimental rheometry data. One of the most interesting characteristics of the GPR is the capability of returning as output a measure of the quality of the regression. With such information, it is possible to make an evaluation of the training process and consider an enlargement of the training dataset. The model itself can suggest where the regression has a lower quality and help the user identify the optimal values for the next experimental samples. Moreover, the computational performance of the GPR model is evaluated by comparison with the computational times required by the power law model. While the GPR unnecessarily increases the computational costs when the rheogram can be accurately interpolated with a functional form, this can be used as an initial test for the approach. The next step would be creating a multidimensional GPR model in which the viscosity is predicted depending on more input variables, e.g. the strain rate, the composition, and the temperature. In such a case the increased computational cost would be compensated by the complexity of obtaining a multivariate function that can accurately represent the viscosity variations.

The results obtained from the CFD simulation are validated against the results of the experimental setup for the same unbaffled stirred tank. From a qualitative point of view, the trend of the power number with respect to the Reynolds number from the simulations agrees with the experimental one. A quantification of the difference between the computational results and the experimental one can be done through the average relative error, which is about 25%.

In conclusion, computational studies can be a precious aid to the experiment once the desired level of accuracy of the model is achieved. The CFD simulations can be useful to predict integral variables related to the power consumption, but also to analyze more in detail the features of the flow that are not accessible experimentally. On the other hand, the GPR remains a promising tool, under the conditions of a satisfactory training step and a model complex enough to justify the increase in computational costs.

[1] Pasquino et al., Soft Matter, 2019,15, 1396-1404

83a Further Analysis of a Simulation of an Axial-Flow Gas Induction Impeller

Eric E. Janz, Kevin Myers, Panagiotis Athanasiou

This is an extension of the work presented at the North American Mixing Forum's Mixing XXIII meeting in the summer of 2024.

Hoffman et al. (2022) presented performance data for a gas induction impeller whose design differed from those previously discussed in the literature in two important ways. First, the impeller generates axial flow rather than radial, and second, the gas is induced through large openings in the impeller rather than small holes. The second difference leads to the induced gas forming bubbles that are significantly smaller than the openings in the impeller as opposed to forming bubbles that are similar in size to the openings at which they are generated.

Laboratory scale experimentation for the design, development and optimization of this class of impellers is more challenging than other types of gas handling impellers. Typical methods for measuring power can be impacted by the gas measuring method (because seals are required for many of the techniques) and methods for measuring the gas flow can result in a pressure drop in the system that would not be present in normal operation (this is even true using most non-contact measurement methods). The problem of measurement and validation becomes even more difficult at the production scale. These types of impellers are typically used with gasses that that are dangerous (flammable – e.g. hydrogen) or corrosive (e.g., fluorine, chlorine, bromine), and methods for measuring the gas flowrates in such setups are prohibitively expensive. In these cases, performance can be inferred from other measurements (e.g., reaction rates, concentration response). A validated computational simulation method will allow for faster optimization of these impellers and assist in the design of industrial scale systems.

Simulation of multiphase flows is inherently complex. Many gas-liquid simulations use Lagrangian particles to simulate gas bubbles in the system. This is a useful simplification of the physics, most notably when the gas entering the systems is already in a bubble form (e.g. a ring sparger or sintered sparger). This allows break-up and coalescence kernels to be used for bubble-bubble and bubble-fluid interactions. For gas induction simulations, the Lagrangian representation alone is insufficient to model the system. An immiscible fluid model is necessary because the gas phase is continuous until it exits the impeller. Once the gas exits the impeller, the simulation resolution must be such that the bubbles of gas can be resolved. This quickly becomes computationally intractable because bubble sizes can become quite small compared to the tank size. To address this issue, a conversion method from the Eulerian immiscible fluid model to a Lagrangian particle can be used. This allows simulation of the induction process and bubble transport be accomplished.

In this study, M-Star CFD will be used to simulate gas induction at two separate scales, a 0.597m (23.5 inch) vessel and a 1.52m (60 inch) vessel. Parameters to be investigated include ungassed power draw, power draw while inducing gas, and induced gas flow rate, all of which will be compared to experimental data. Given that the impeller of interest pumps axially, the effect of pumping mode (i.e. – down-pumping or up-pumping) on these parameters is also of interest.

References:
Hoffman, Shannon M., Eric E. Janz, Kevin J. Myers, and Nicholas A. Brown. 2022. "Performance Characteristics of an Axial-Flow Gas Induction Impeller." The Canadian Journal of Chemical Engineering 101 (3): 1371–86. https://doi.org/10.1002/cjce.24556

145d Extension of Mean Age Theory to Transient Inlet Velocity Conditions

John White, R. Eric Berson

Mean age theory introduces a redefined perspective on time as a passive scalar variable within the advection-diffusion equation. This allows for the analysis of traditionally time-based variables, such as mean residence time, while utilizing a steady-state solution. Unlike conventional methods that demand time-consuming and computationally intensive transient solutions for the advection-diffusion equation, mean age theory offers a more efficient alternative. Even when a steady-state solution is desired, conventional methods necessitate transient calculations to model time-dependent tracer behavior. Computational fluid dynamic (CFD) solutions incorporating mean age theory offer additional benefits. They not only provide spatial and temporal resolution within the flow field but also offer insights beyond conventional velocity vector and contour plots. Moreover, they offer solutions in scenarios where traditional experimental measurements are challenging or impractical.

The conventional form of the mean age transport equation assumes a time-invariant velocity input boundary condition. In our work here, we derive a variation of the scalar transport equation that accommodates a time-dependent velocity input at the boundary. Beginning from the same advection-diffusion equation, we derive an expression analogous to the well-studied steady mean age transport equation. However, our derivation includes an additional term incorporating [d(velocity)/dt] to account for the time-dependent velocity. Consequently, our output provides mean age as a function of time over the period defined by the velocity input function. Notably, the average age over this period differs from that obtained when assuming a constant velocity equal to the average of the velocity function. Specifically, the time-averaged age at the exit combines the mean age derived from a constant velocity input with the contribution from the additional term containing the input velocity function, d(velocity)/dt.

To illustrate our approach, we apply it to the study of blood flow through a stenosed or constricted artery segment, where blood flow follows a pulse function rather than a constant velocity. The average mean age at the exit over the duration of one pulse differs from the mean age when the velocity input is defined as a constant value using the average velocity of the pulse. For instance, the average mean age with the pulse input was 0.080 secs, compared to 0.053 secs for the constant velocity input (refer to Figure 1 for pulse function and constant velocity). This disparity in mean age is attributed to the acceleration and deceleration of the blood, particularly affecting recirculation/holdup zones past the stenosis (see Figure 2), resulting in prolonged path lengths and greater residence times or mean age. The delay in a holdup region is more pronounced for pulsing velocity compared to constant velocity.

By computing mean age for each time step in the transient solution, our approach significantly reduces computational effort compared to computing mean residence time from time-dependent tracer analysis.

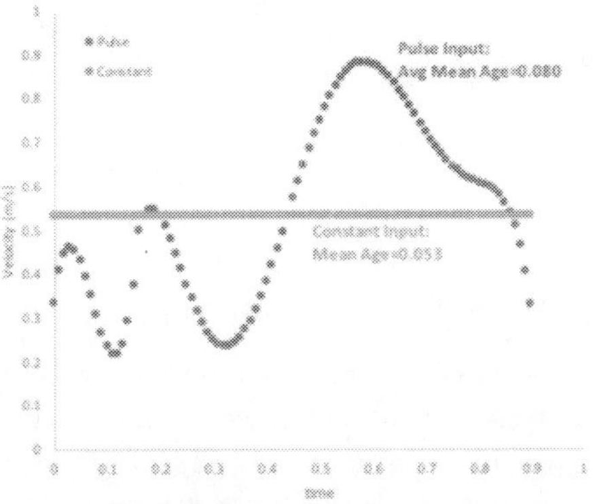

Figure 1. Comparison of mean exit age computed values for arterial flow between a pulsing velocity input (blue) and a constant input (orange)

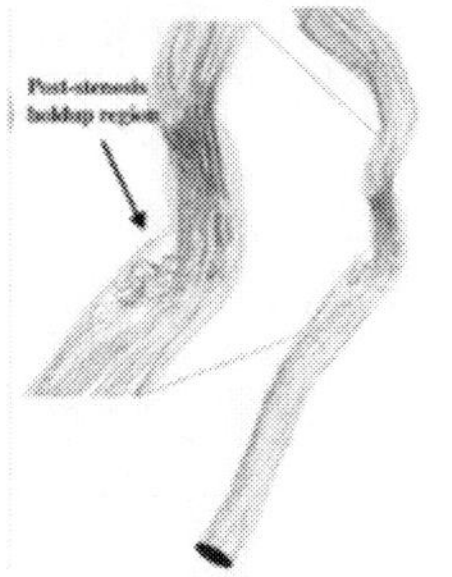

Figure 2. Illustration of recirculation region resulting from the abrupt contraction and expansion of flow volume caused by plaque build-up.

145e Scale Down Studies for Complex Fluids Compounding Using Multiphase Simulation.

Mandakini Padhi, Amarnath Reddy Ganji, Swarali Sandeep Kadam, V Selvakumari, Srichakravarthy Narasimharaghavan, Jean-Sebastian Simard, Atul Awasthi

The study investigates the application of multiphase simulation techniques for conducting scale-down studies of complex fluid mixing. Complex fluids, such as emulsions, suspensions, and foams, play a crucial role in various industrial processes, including pharmaceuticals, food, and oil and gas. Understanding their behaviour at different scales is vital for process optimization, cost reduction, and product quality improvement. However, experimental scale-down studies can be challenging and costly due to the intricate nature of these fluids. In this research, we propose the utilization of multiphase simulations as a cost-effective alternative to experimental scale-down studies. By employing advanced computational fluid dynamics (CFD) models, we can accurately capture the complex dynamics, including rheological properties of the fluids. The results show that multiphase simulations provide valuable insights for the identification of key process parameters and optimization of scale-down studies. Overall, this study demonstrates the potential of multiphase simulations as a powerful tool for scale-down studies of complex fluids, offering significant benefits in terms of cost, time, and process understanding.

145g CFD analysis of shear rate distribution and the Metzner-Otto constant with different concentrations of shear-thinning fluid in a stirred vessel

Mehak Jain, Ryuta Misumi

Background: In the chemical industries, different kinds of impellers are frequently employed for the mixing of pseudo- plastic fluids to resolve inadequate mixing. Power consumption and mixing time plays an important role in the evaluation of mixing quality. For accurate determination of power consumption and shear rate; Metzner-Otto concept (Metzner and Otto 1957) is used widely to optimize the mixing process for non-Newtonian shear-thinning fluids. By using Computational Fluid Dynamics (CFD) software it is possible to compute the shear rate distribution in a stirred tank.

Method and Materials: In this study, power consumption and average shear rate were investigated with Newtonian and different concentrations of non-Newtonian shear-thinning fluid in the stirred tank vessel stirred by six-blade paddle impeller, to understand the effect of rheological parameters of non-Newtonian shear-thinning fluid on Metzner-Otto constant using multi-purpose CFD software. For stirring liquid starch syrup (η = 6.7 Pa·s) and polyglycerin (η = 26.7 Pa·s) were used as Newtonian fluids, and an aqueous solution of hydroxyethyl cellulose (HEC) was used as pseudoplastic fluid. HEC concentration varied from 1.75 wt.% to 3 wt.%. Rheometer (ONRH–1) and Carreau model were used to calculate the rheological parameters of the different concentrations of HEC, and the flow behavior index n was in the range of 0.62 – 0.48.

Experimental results: The results obtained from previous study (Jain and Misumi 2024) suggest that power number decreased linearly with increasing Reynold number in the laminar flow regime, and using the Metzner–Otto method, all power curves for the shear-thinning fluids coincide with those of the Newtonian fluids. Double power needs to rotate the double-stage impeller (2S) as compared to single-stage impeller (1S). Further, Metzner-Otto constant *K*s,exp was found to be dependent on power index n for non-Newtonian highly shear-thinning fluids for 1S and *K*s,exp value is same for 1S and 2S.

CFD analysis: To check the applicability of the experimental results i.e. to understand the dependence of *K*s on n for single-stage impeller, power dissipation and shear rate distribution was analyzed by using multi-purpose CFD software. For the CFD analysis, a similar geometry of the experimental stirred tank with different impellers was constructed. To validate the CFD, power results of CFD were compared with the experimental power results. Direct numerical simulation (DNS) results were in excellent agreement with the experimental results of power consumption for both Newtonian and non-Newtonian shear-thinning fluids. Local shear rate distribution for Newtonian and shear-thinning fluids was analyzed at different impeller speeds. From the local shear rate distribution, it is evident that local shear rate changes with change in the shear-thinning behavior of non-Newtonian shear-thinning fluids at the same impeller speed and this could be the reason of the dependence of *K*s on n. DNS results of double-stage impeller (2S) shows that power dissipation is double for 2S as compared to 1S. With the help of MATLAB, volumetric flow region was calculated for 1S and 2S at different shear rates and found out that volumetric flow region is double for 2S as compared to 1S at the same average shear rate. As the average shear rate is same for 1S and 2S, *K*s is also same for 1S and 2S.

References:

1. Metzner, A. B., & Otto, R. E. (1957). AIChE Journal, 3(1), 3-10.
2. Jain, M., and R. Misumi. (2024). Journal of Chemical Engineering of Japan, 57 (1): 2387459.

Acknowledgment:

This study was financially supported by the JSPS KAKENHI (Grant No. 22K04799).

213a Application of a Novel GPU Native CFD Solver to Examine and Optimize Micro-Mixing Devices for Nano-Lipid Production

Luke Munholand, Hossam Metwally

Lipid nanoparticles (LNP) may be produced using mixing devices such as Combined Impinging Jet Mixers (CIJM) or many styles of microfluidic reactors. In all devices, optimal micromixing is key. Computational Fluid Dynamics (CFD) is an effective tool for study and optimization any mixing device. A next generation GPU native CFD code has been used to characterize micro-reactor performance for the production of LNPs. Reference geometry is presented with boundary conditions, physics models and methods, and computational grid convergence. Key results including mixing parameters such as velocity, strain rate, turbulence and species concentration are presented. Comparison to published lab results is included. Well validated CFD is an important first step to attaining an optimized production of properly sized, mono-disperse, well encapsulated LNPs.

213g Evaluation and selection of mixing systems for sterile drug products with "predict first" smart process engineering, modeling and PAT approaches

Parind Desai, Mark Palmer, Meghan McDermott, Warren Eagles, Zahra Awan, Zoe Zhang, Alexis Venere, Igor Rusanov, Himanshu Bhattacharjee

Mixing operation could be challenging to scale-up for floating and settling particles for sterile drug products. This presentation will cover holistic smart process engineering approaches researchers can adopt while selecting and/or evaluating mixing systems. These approaches are underpinned by API sparing surrogates' evaluation, "predict first" mechanistic modelling, engineering calculations and scaled-down experimentation. In this presentation, we will a) understand and acknowledge complexities involved while selecting and/or evaluating commercial scale mixing systems for Steriles product development, b) apprehend the importance of drug substance sparing surrogates' evaluation, scaled-down experimentation, "predict first" CFD - mechanistic modelling and engineering calculations in compressed timelines of drug product development, c) learn the need of different mixing systems based on the properties of drug substance and drug product.

213c Validation of a Novel Stress Blended Eddy Simulation (SBES) to Predict Hydrodynamics of Stirred Tank Flows in the Transitional Flow Regime.

Georgina Wadsley, David F. Fletcher, Andrew Ingram, Joelle Aubin, Waldo Rosales, Mark J. H. Simmons

Validated Computational Fluid Dynamics (CFD) simulations when applied to industrial flows are powerful predictive tools that facilitate rapid modification and improvement of existing processes. The complex rheological properties of home and personal care products typically evolve during manufacture, as functional microstructures build. This typically causes the apparent viscosity to increase, thus causing the flow to move from turbulent, through transitional, to the laminar flow regime in the mixer. In the transitional flow regime, hydrodynamic instabilities create unsteady flows that are difficult to predict using existing laminar or turbulence closure models that were developed for flows with high Reynolds number turbulence [1]. Hence, further advancements in the effective application of CFD to simulate mixing in the transitional flow regime are required.

Focusing on transitional regime flows in a stirred tank with a hydrofoil impeller [1], a dynamic turbulence closure model – stress blended eddy simulation (SBES) – has been evaluated alongside a generalised k-ω shear stress transport model (GEKO), through comparison with experimental 2D Particle Image Velocimetry (PIV) and torque measurements (Fig 1). Unsteady flows of water, Newtonian glycerol solutions (40–100 wt%), and non-Newtonian CMC solutions (0.1–1.5 wt%) were measured over a wide range of values of Reynolds number, Re, from 8 to 25,000, spanning laminar to turbulent flow via the transitional regime. For both the Newtonian and non-Newtonian fluids, two flow scenarios were investigated for each value of Re, scaled by changing tip speed (V_{tip}) according to apparent fluid viscosity. Using Ansys Fluent (version 2024R1), transient CFD simulations were built for selected cases of interest with a ~2.4M element mesh, giving mesh independence. Proper orthogonal decomposition (POD) of PIV data and periodic averaging of CFD allowed exploration of the impact of pseudo-turbulence arising from impeller blade and baffle periodicity.

PIV results indicated that, for constant viscosity, increasing Re augmented both the magnitude and fluctuations of velocity. However, when normalised by V_{tip}, velocity initially decreased with increasing Re for laminar flows before increasing through the transitional regime, then achieving independence of Re in fully turbulent flow. Normalised velocity fluctuations increased with Re across all flow regimes. The values of Re for which transition between these laminar and turbulent behaviours occur, and thus transitional regime boundaries, change depending on fluid viscosity. The transition to fully turbulent flow occurs at decreasing values of Re for increasing apparent fluid viscosity. Two cases with similar transitional Reynolds number (Re = 697 and 673) but different viscosity ($\mu = 0.051$ and 0.012 mPa s) scaled by agitation speed ($V_{tip} = 0.98$ and 0.23 m s^{-1}, respectively) exhibit almost similar flow hydrodynamics. Displaying similar normalised velocities (Fig 2a) and fluctuations (Fig 2b), which would be expected more in the fully turbulent regime.

Both GEKO and SBES closure models have been validated for torque prediction. However, SBES gives much better agreement between values of impeller and wall torque in the transitional regime. When comparing with PIV data (Fig. 3), SBES outperforms GEKO for predicting flow hydrodynamics, at the cost of additional computational requirement. Machine learning is to be implemented with the aim of modifying the GEKO constants to give a more accurate prediction, as given by SBES, providing a more industrially relevant model.

References

1. Mendoza, B. Lopes Bañales, E. Cid, C. Xuereb, M. Poux, D. F. Fletcher, J. Aubin, *Chemical Engineering Research and Design*, 132 (2018) 865-880.
2. Menter, A. Hüppe, A. Matyushenko, D. Kolmogorov, *Applied Science*, 11 (2021) 2459-2473.

Acknowledgments

GW is supported by the EPSRC Centre for Doctoral Training in Formulation Engineering (EP/S023070/1) at the University of Birmingham and Unilever Port Sunlight, UK.

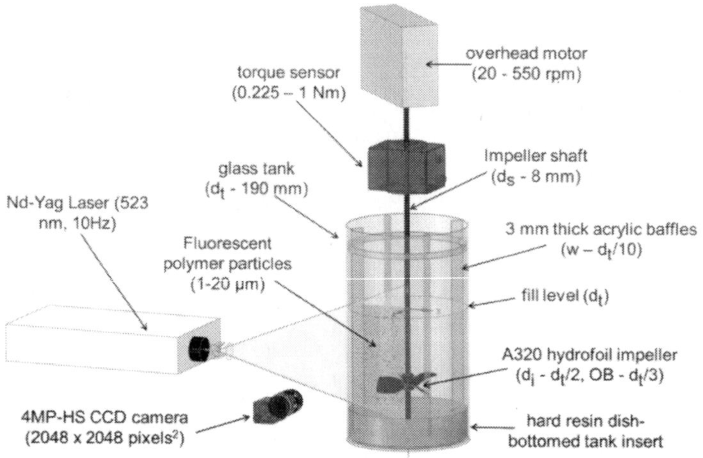

Fig 1: Schematic of experimental set-up for PIV and torque measurement.

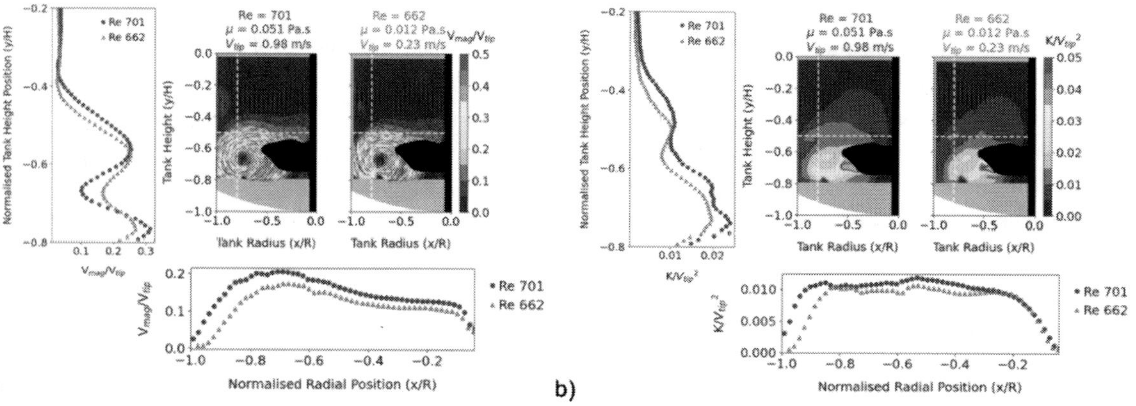

Fig 2: Contour plots of PIV data with radial and axial lines (−0.8R from impeller and 0.3H from tank bottom, respectively) of a) normalised velocity magnitude and b) normalised velocity fluctuations for two cases at Re = 697 and 673 (left/blue - V_{tip} = 0.98 m s^{-1}, μ = 0.051 Pa s, right/red - V_{tip} = 0.23 m s^{-1}, μ = 0.012 Pa s).

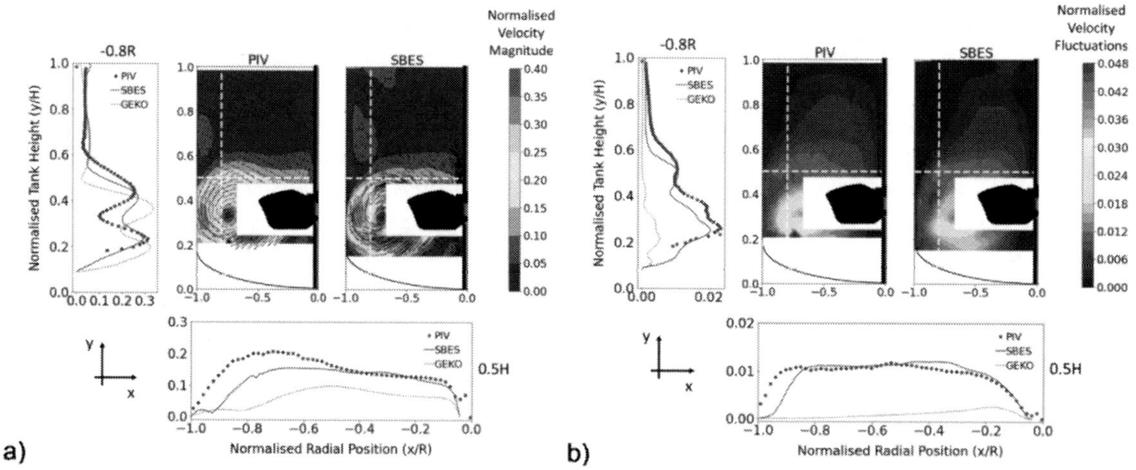

Fig 3: Contours comparing PIV (left/markers) data with SBES (right/solid line) and GEKO (dotted line) predictions with radial and axial lines (−0.8R from impeller and 0.3H from tank bottom, respectively) of a) normalised velocity magnitude b) normalised velocity fluctuations for a single case (Re = 697, V_{tip} = 0.98 m s^{-1}, μ = 0.051 Pa s).

213d Combining Machine Learning and CFD Simulation to Evaluate Design and Operating Conditions Faster

Thomas Eppinger, Justin Hodges, Ravindra Aglave, Arman Raoufi

For the process industry, the digital tools for design and operation of equipment is ripe for supplementation with machine learning. Design engineers can benefit from the constant stream of innovations in the ML community when productizing new equipment, while operators of the equipment can be enhanced in productivity from on-demand information when these ML models are deployed for their easy access. There are several themes by which machine learning can cause a material impact: augmented user experiences, optimization, better utilization of physical measurement, and time savings via faster insights.

Utilization of a limited number of experiments and complimenting them with simulated data can result in significant time savings and valuable insights into engineering processes. Specifically, how simulation experts, a widely used engineering method, can realize large time savings when employing ML models in their sensitivity and design exploration, which can help reduce time to market. Further, how physical measurement of in-operation equipment can be combined with simulations with as input to (ML powered) executive digital twins (xDTs) to provide the operator with simulation-level detail of performance from limited discrete physical measurement points.

In this study, we perform multiple batches of computational fluid dynamics (CFD) simulations to characterize the performance of a 20 cubicmeter aerated stirred bioreactor with respect to changes in the geometric design as well as the operating conditions in terms of stirrer speed and aeration rate. The final goal is to optimize the reactor in terms of energy consumption and productivity. More significantly, we then explore a series of three independent scenarios whereby unique ML-based ROMs are created and analyzed. First, we conduct an optimization study in a hybrid modality: machine learning models are used when possible, during the optimization study in conjunction with traditional simulation for the design points. Second, we create and leverage ML-based ROMs for adjacent design characterization akin to typical robustness, reliability, and sensitivity analysis. Finally, we showcase the time and hardware savings when using AI-based (dynamic) sampling for creating simulation datasets from user-specified objectives when only a small number of simulations are budgeted to run.

213h A Unified Framework for Total Variation Regularized Optimization in Fluid Dynamics and Related Physical Systems

Varsha Gupta

An optimization framework is presented for minimizing the energy functional developed around a generalized equation governing physical systems such as fluid dynamics, particle transport, phase transition, and other related systems. The convexity of the energy functional is investigated to derive the necessary conditions for a smooth and global optimum solution. Furthermore, the Total Variation (TV) regularization term is introduced to gain insights into the solution space and convergence analysis of convection- dominated problems. The framework optimizes energy functionals associated with fluid systems, offering improved convergence and smoothness in solutions where non-linear interactions—such as those seen in turbulent flow or agitated systems—lead to complex behavior. By applying TV regularization, the optimization reduces noise and sharp gradients in the solution space, critical in accurately modeling and optimizing mixing efficiency in industrial settings, such as stirred tank reactors and turbulent flow mixers. Through applications to the Navier-Stokes equations, we demonstrate how the framework can be used to model and optimize mixing patterns, helping to understand better fluid motion in scenarios where convection plays key roles.

83c Simulation of a Radial Flow Gas Induction Agitator System

Eric E. Janz, Navraj Hanspal

Hoffman *et al.* (2022) presented performance data for a gas induction impeller. This data was collected over three different laboratory scales (tanks of 0.4445 m, 0.5969 m and 1.524 m) and used as part of the validation and verification effort with M-Star CFD that will/was presented at the joint NAMF-Mixing XXVIII and ISMIP 11 conference in June 2024. (Janz, Myers, Athanasiou, 2024).

Ultimately, the CFD method's usefulness is examining the performance of industrial scale applications where measurement of the actual flow rates or just induced speeds are much more difficult than at the lab and pilot scales. Not only are the process gases regularly hazardous (e.g. hydrogen, chlorine, bromide), the processes are typically carried out under significant pressure. Using a viewing port on vessel top, an investigator may be able to identify when gas starts to evolve from the liquid surface, or a relative intensity of the gas being evolved but not quantitative values.

In this study, an industrial application of a radial flow gas induction impeller is analyzed via CFD using M-Star CFD. From plant observations, the minimum induction speed for this process is known but other details are limited. Using method from the Janz, *et al.* an immiscible fluid model will be used with a Eulerian immiscible fluid to Lagrangian particle conversion method to allow simulation of the induction process and bubble transport in a reasonable time from (hours to days on dedicated desktop hardware).

The objectives of this study are to:

- Accurately predict the just induced speed based on the plant operating parameters.
- Develop a correlation for the gas induction speed given liquid height and fluid conditions.
- Develop a correlation for gas flow rate after the agitator speed has exceeded the just induced speed.

References:

1. Hoffman, Shannon M., Eric E. Janz, Kevin J. Myers, and Nicholas A. Brown. 2022. *Performance Characteristics of an Axial-Flow Gas Induction Impeller.* The Canadian Journal of Chemical Engineering 101 (3): 1371–86. https://doi.org/10.1002/cjce.24556
2. Janz, Eric. E., Kevin. J Myers, & Panagiotis Athanasiou, P. *Simulation of an Axial-Flow Gas Induction Impeller.*Mixing XXVII.

1a Eccentric Mixing with a Maxblend Impeller

Kazuhiko Nishi, Takuma Kato

Various kinds of large impellers, such as FULLZONE (Kobelco Eco-Solutions Co., Ltd.), Super-Mix MR205 (Satake Chemical Equipment Mfg Co., Ltd.), Hi-F mixer (Soken Tecnix Co., Ltd.) and MAXBLEND (Sumitomo Heavy Industries Co., Ltd.), have been developed in Japan. Since these impellers have a high mixing performance over a wide range of viscosities, they are used in mixing, dispersion, reaction and polymerization processes. Recently, their use in the food and pharmaceutical industries is being considered. For agitation in the turbulent region, these large impellers are usually used with baffles to promote mixing. However, baffles often cause problems for washing and sterilization. Furthermore, in the laminar region, baffles are not effective for mixing, and in fact, they often obstruct mixing. Eccentric mixing is one of the traditional methods of promoting mixing in a vessel without baffles. An eccentrically located impeller generates a vertical flow, which contributes to mixing, without baffles. If a large impeller is used at an eccentric position, it is expected that the high performance of the large impeller can be combined with the advantages of an eccentric mixer. In this study, a MAXBLEND impeller was investigated as an example of a large impeller. The power consumption and mixing time of a MAXBLEND impeller were measured under various eccentric conditions. Furthermore, the mixing performance based on the power consumption of eccentric mixing with a MAXBLEND impeller was investigated.

When eccentric mixing is used industrially, we should be concerned about the horizontal load to an agitating shaft. It is expected that the average torque and horizontal load on agitating shaft are larger than in the concentric mixing without baffles. Since these values fluctuate with the rotation of the impeller, the instantaneous maximum value is still larger. The large, fluctuating torque and horizontal load can cause serious problems, such as the falling off of the impeller or the breakage of the shaft, motor, mechanical seal or gearbox. It is, therefore, important to understand the relation between these values and the impeller rotational speed when designing the mixing equipment and determining the operating conditions. In this study, the torque and horizontal load were measured in eccentric mixing using a MAXBLEND impeller, as an example of a large impeller, at various impeller rotational speeds and under various eccentric conditions in a turbulent state. The average torque and standard deviation, corresponding to the amplitude of fluctuation were calculated, and the cause of the fluctuation was investigated by FFT analysis.

Power consumption and mixing time of MAXBLEND eccentric mixing

The power consumption in neat glycerol was proportional to the square of the impeller rotational speed. It was presumed that the flow state was laminar. The change in power consumption with the eccentric length was small. The power consumption at $L_E = 0.04$ m was slightly larger than that of concentric mixing with baffles. The power consumption in water was smaller than that in glycerol, and it was proportional to the cube of the impeller rotational speed. This indicates that flow was in a turbulent state. When the eccentric length increased, the power consumption increased. The power consumption at L_E=0.04 m was almost the same as that of baffled mixing. In **Figure 1**, the relation between the impeller rotational speed and the power consumption was shown by the correlation of the Reynolds number, Re (= $\rho n d^2/\mu$) and power number, N_p (= $P/\rho n^3 d^5$). In the laminar region, N_p of eccentric mixing decreased linearly with increasing Re on a logarithmic graph. The slope of N_p against Re was approximately −1. When L_E was large, N_p was large. The difference between N_p of eccentric mixing and N_p of concentric mixing with and without baffles was small. It was interesting that the power consumption at $L_E = 0.04$ m was larger than that of concentric mixing with baffles. In the turbulent region ($Re > 2000$), N_p at $L_E = 0.01$ and 0.02 m decreased linearly with increasing Re. This behavior was similar to that of concentric mixing without baffles. On the other hand, the values of N_p at $L_E = 0.03$ and 0.04 m were constant. This behavior was similar to that of concentric mixing with baffles. Since N_p at $L_E = 0.03$ and 0.04 m was independent of Re, it was presumed from the power consumption that a turbulent state was established throughout the whole of the vessel. In eccentric mixing, since a vortex, which was very large in concentric mixing without baffles, was not generated near the mixer shaft, it was presumed that the generation of a cylindrical rotating zone was inhibited. In this region, N_p of eccentric mixing was clearly larger than that of concentric mixing without baffles ($L_E = 0$). It was, thus, confirmed that a large amount of energy can be supplied to the mixing liquid in eccentric mixing.

The mixing time was considered a non-dimensional quantity, given by $n\theta_M$. The relation between $n\theta_M$ and Re is shown in **Figure 2**. In the region where $Re < 2000$, when Re increased, $n\theta_M$ decreased. In this region, which was considered to be in a laminar or a transition state, the value of $n\theta_M$ of eccentric mixing was smaller than that of concentric mixing with and without baffles. The eccentric impeller generated asymmetrical flow, which promoted mixing. On the other hand, in the region where $Re > 2000$, the value of $n\theta_M$ was almost constant for each value of L_E. This was characteristic of a turbulent state. In this region, the value of $n\theta_M$ of eccentric mixing was clearly smaller than that of concentric mixing without baffles. The $n\theta_M$ at $L_E = 0.01–0.03$ m of eccentric mixing was larger than that of concentric mixing with baffles. However, the value of $n\theta_M$ at $L_E = 0.04$ m was almost the same as that of concentric mixing with baffles. Turbulent diffusion dominated $n\theta_M$ during mixing in a turbulent state. It was predicted that the turbulence intensity of eccentric mixing would be lower than that of concentric mixing with baffles.

Torque and horizontal load of MAXBLEND eccentric mixing

The torque of eccentric mixing was fluctuated periodically. When the eccentric length L_E, which was distance from the center of vessel to impeller shaft, increased, the average torque increased. FFT analysis showed that the frequency of torque was twice the impeller rotational speed, n, and that an amplitude of fluctuation became large by the increase in L_E. According to CFD, it was found that a large drag force was generated on the blade near the center of the vessel, and the torque had maximum peaks. In eccentric mixing with a MAXBLEND impeller, which has two blades, and since the blades approach the center of vessel twice during each rotation, the torque fluctuates with a frequency of $2\,n$ [Hz]. In mixing with water, the average of torque, T_{ave} was almost proportional to the square of n. This is typical behavior in turbulent mixing. T_{ave} increased exponentially with L_E, and the gradients of each n were almost the same. When L_E increases, the speed of blade through the vessel center and working length of drag force increase. Therefore, the torque as a moment increased. The relation between the standard deviation of the torque, T_{std}, which corresponds to the amplitude of the torque fluctuation, increased with the increase in eccentric length. T_{std}, particularly increased greatly in larger eccentric length. When T_{std} was compared with T_{ave}, each T_{std} was 15% or less of T_{ave} at the same eccentric length and impeller rotational speed. In the design of a shaft diameter in an eccentric mixer using a MAXBLEND impeller, it is necessary to prepare larger torque than T_{ave} based on T_{std}.

The load in normal – direction (y) to the blade, F_y was fluctuating between plus and minus periodically, and their average was almost zero in both cases. The period of F_y was the same as the impeller rotational speed. The strain gauges for measuring F_y were affixed to the rotational shaft. The period of F_y means that forces in both, positive and negative directions, were loaded on the impeller during each rotation. This corresponded with the blade twice passing the center vessel wall during each rotation. The amplitude of F_y increased in proportion to the square of n. This may indicate that F_y is caused by an unbalanced load on the two blades of MAXBLEND. F_y increased exponentially with L_E, and the gradients of each n were almost the same.

Material strength analysis was performed based on measuring torque and horizontal load of MAXBLEND eccentric mixing. As a result, it was found that the horizontal load has a large influence on the principal stress and stress concentration in the agitating shaft, and that value and period of horizontal load are important for equipment design and estimation of fatigue strength.

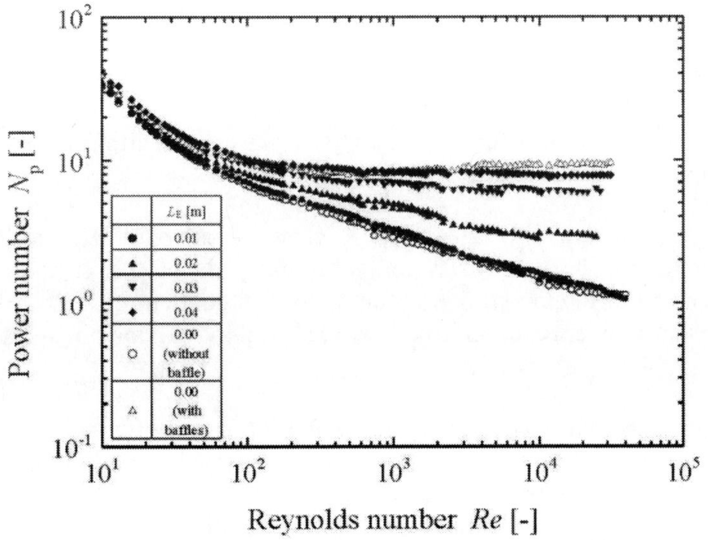

Figure 1 Relation between N_p and Re

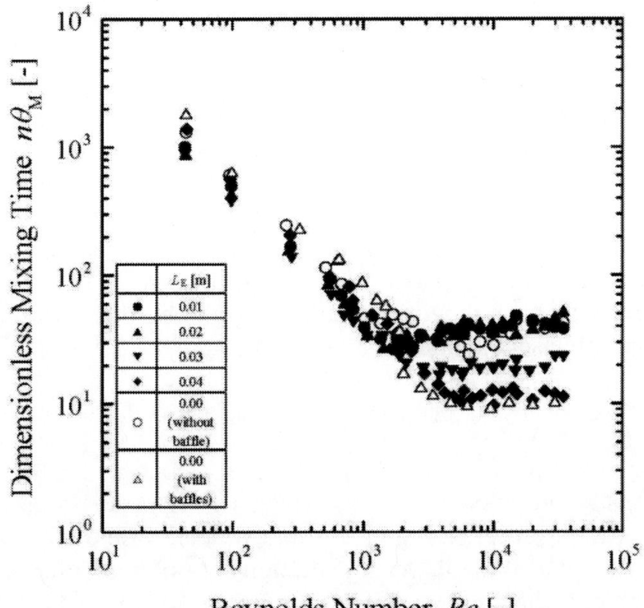

Figure 2 Relation between $n\theta_M$ and Re

1b Characterization of Specialized Mixing Elements Applied in a Novel Single-Screw Extrusion Reactor Designed to Process and Enhance Mixing in Highly Viscous Reactive Systems.

Jonas Loncke, Leen Braeken, Simon Kuhn, Leen C.J. Thomassen

Introduction

Current flow processes generally maintain dilute reagent conditions to avoid fouling and blockages in the reactor. Consequently, there is potential to advance towards more sustainable flow processes by reducing or even fully eliminating the amount of solvent used in these processes. Nevertheless, decreasing the solvent content from reactive systems often increases its viscosity, resulting in reduced control over the molar flow rates, and heat- and mass-transfer limitations inside the reactor [1].

Extruders, like single screw extruders (SSE) and twin screw extruders (TSE), are highly advantageous for the transportation and mixing of high viscosity and solid materials. However, implementing liquid/liquid and liquid/solid chemical synthesis in extrusion reactors is still challenging, as it often results in the formation of byproducts and lower reaction kinetics compared to reactions in solution due to a reduced mixing efficiency [2].

In literature, Mixing Elements (ME's) were previously applied in polymer chemistry to enhance polymer blending and processability inside extruders by inducing shear stresses and elongational flow [3]. These literature ME's are roughly divided into two categories based on their geometry: Pin mixers (Normal and Pineapple) and Slotted Flight Mixers (Dulmage and Saxton). Figure 1 showcases examples of each category. However, experimental micromixing characterization of these ME's remains scarce in literature with the exception of the research performed by Rożeń et al. [4], [5], [6]. Furthermore, this micromixing characterization was exclusively performed in aqueous media. Hence, there is no unambiguously conclusion on which ME-geometry provides the most optimal macro- and micromixing performance at high viscosities.

In conclusion, solvent-reduced synthesis in extrusion processes faces challenges related to mixing due to the high viscosity of the reaction mixture. Therefore, these challenges will be tackled in this research by investigating macro- and micromixing properties of four distinct mixing elements applied in a specialized single-screw extrusion (SSE) reactor platform. The SSE-reactor platform, visualized in Figure 2, was designed and manufactured in-house. Thus, the primary goal of this research is to determine the most optimal ME-geometry to enhance mixing performance in high viscosity media. In addition, experimental characterization techniques to quantify the macro- and micromixing performance in extruders will be adapted for applications in high viscosity media.

Methodology

The macro- and micromixing behavior in the SSE platform equipped with ME's is experimentally characterized at several constant viscosities ranging between 1 mPa.s to 100 Pa.s. Mixtures with viscosities up to 1 Pa.s will be obtained using solutions of glycerol in water, while mixtures with viscosities ranging between 1 Pa.s and 100 Pa.s will be obtained using solutions of polyvinylalcohol in water. Furthermore, ME-geometry (Pin, Pineapple, Dulmage, Saxton) and screw configuration (T-ME-T, ME-ME-T, ME-T-ME with T = transport element and ME = mixing element, Figure 3) and screw rotational speed (0 to 400 RPM) are studied.

Macromixing refers to the mixing that occurs at macro scales which affects the distribution of reactants and heat within the reactor [7]. Residence time distribution (RTD) is applied to quantify macromixing, more specifically axial dispersion. A step signal of an inert tracer or salt is injected into the feed of the SSE-ME platform. The concentration of salt is measured at the exit of the reactor by an inline conductometer. The macromixing efficiency can be quantified using the mean residence time (), variance

() and the dispersion number (D) where a larger value of the dispersion number represents a larger amount of axial dispersion.

On the other hand, micromixing occurs at the molecular level, where reactants come into contact and interact to initiate chemical reactions influencing the kinetics, selectivity and yield of the reaction [7]. The well-known Villermaux-Dushman protocol will be employed to characterize micromixing. This technique employs two competing reactions: a neutralization reaction (R1) and a redox reaction (R2), both of which consume acid-protons ( ) [8]. First of all, the protocol is upscaled to higher viscosities in batch and benchmarked against the literature reported (batch) protocols with glycerol and sucrose. Next, the upscaled protocol is applied in the SSE-ME platform, equipped with an inline UV-VIS spectrophotometer where the effect of the aforementioned parameters on the micromixing efficiency is studied.

Results and Discussion

Firstly, the effect of the screw rotational speed and feed flowrate on the axial dispersion or macromixing performance was studied (Figure 4). When no rotational speed is applied, the mixing elements behave as static mixers and the transport elements provide no additional forward axial movement of the fluid. When increasing the rotational speed from 0 to 100RPM, there is a decrease in axial dispersion due to the forward axial movement of the fluid imposed by the rotation of the helical transport screw. However, when a rotational speed between 100 and 400 RPM is applied to the mixing screw, the axial dispersion increases when increasing the rotational speed, indicating the occurrence of backmixing. Furthermore, increasing the viscosity of the feed results in increased axial dispersion and backmixing at the same rotational speed and feed flowrate. Lastly, applying higher feed flowrates reduces the amount of backmixing and induces more plug-flow behavior for all viscosities. These results suggested that the distance between the transport elements and the barrel had to be decreased to prevent fluid moving backwards in the reactor thus decreasing backmixing. Further experiments are being conducted to determine the most optimal ME-type and ME-configuration for macromixing.

In addition, the Villermaux-Dushman protocol was successfully upscaled to a viscosity of 1 Pa.s in a batch reactor. Batch micromixing results indicate that reactor volume, injection time and stirrer speed are important factors influencing micromixing. It was concluded that higher stirring speed are required, for the 1 Pa.s mixtures compared to 1 mPa.s, to achieve the same degree of micromixing. In the succeeding experiments, the protocol will be scaled up further to a viscosity of 10 Pa.s. Once the protocol is fully optimized, it will be applied to study micromixing in the SSE-ME platform.

Conclusion

Macromixing was studied in the SSE-ME platform at different viscosities (0.001 to 1 Pa.s), rotational speeds and feed flowrates. The results suggested design improvements to reduce backmixing in the reactor. Upcoming experiments will involve studying the macromixing performance of other ME-types and ME-configurations.

Furthermore, the Villermaux-Dushman protocol was successfully performed in batch at 0.001 Pa.s and 1 Pa.s. The protocol will be scaled up further to 10 Pa.s before micromixing tests in the SSE-ME platform will be executed.

References

[1] D. Crawford, C. Miskimmin, A. Albadarin, G. Walker, and S. James, "Organic synthesis by Twin Screw Extrusion (TSE): Continuous, scalable and solvent-free," *Green Chem.*, vol. 19, Mar. 2017, doi: 10.1039/C6GC03413F.

[2] D. Crawford, J. Casaban, R. Haydon, N. Giri, T. McNally, and S. L. James, "Synthesis by extrusion: continuous, large-scale preparation of MOFs using little or no solvent," *Chem Sci*, vol. 6, no. 3, pp. 1645–1649, 2015, doi: 10.1039/C4SC03217A.

[3] H. Cheng and I. Manas-Zloczower, "Study of mixing efficiency in kneading discs of co-rotating twin-screw extruders," *Polym Eng Sci*, vol. 37, no. 6, pp. 1082–1090, 1997, doi: https://doi.org/10.1002/pen.11753.

[4] A. Rożeń, R. A. Bakker, and J. Bałdyga, "Application of Parallel Test Reactions to Study Micromixing in a Co-Rotating Twin-Screw Extruder," *10th European Conference on Mixing*, pp. 85–92, Jan. 2000, doi: 10.1016/B978-044450476-0/50012-1.

[5] A. Rozeń, R. A. Bakker, and J. Baldyga, "Application of an integral method to modelling of laminar micromixing," *Chemical Engineering Journal*, vol. 84, no. 3, pp. 413–428, Dec. 2001, doi: 10.1016/S1385-8947(00)00273-4.

[6] A. Rozeń, R. A. Bakker, and J. Bałdyga, "Effect of Operating Parameters and Screw Geometry on Micromixing in a Co-Rotating Twin-Screw Extruder," *Chemical Engineering Research and Design*, vol. 79, no. 8, pp. 938–942, Nov. 2001, doi: 10.1205/02638760152721163.

[7] S. Gobert, S. Kuhn, L. Braeken, and L. Thomassen, "Characterization of Milli- and Microflow Reactors: Mixing Efficiency and Residence Time Distribution," *Org Process Res Dev*, vol. 21, Feb. 2017, doi: 10.1021/acs.oprd.6b00359.

[8] J. M. Commenge and L. Falk, "Villermaux–Dushman protocol for experimental characterization of micromixers," *Chemical Engineering and Processing: Process Intensification*, vol. 50, no. 10, pp. 979–990, Oct. 2011, doi: 10.1016/J.CEP.2011.06.006.

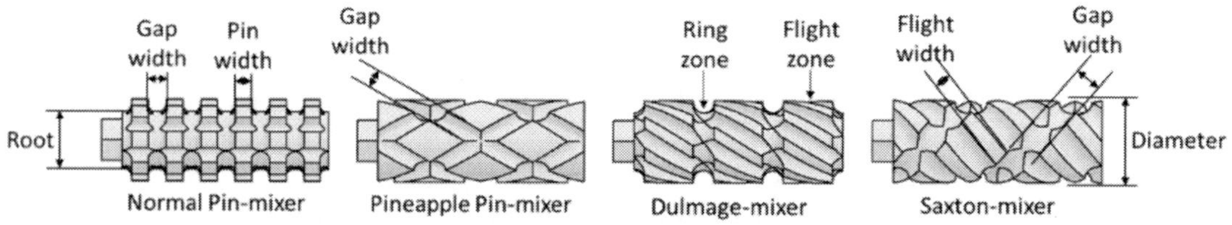

Figure 1: Examples of standard EME's applied in literature.

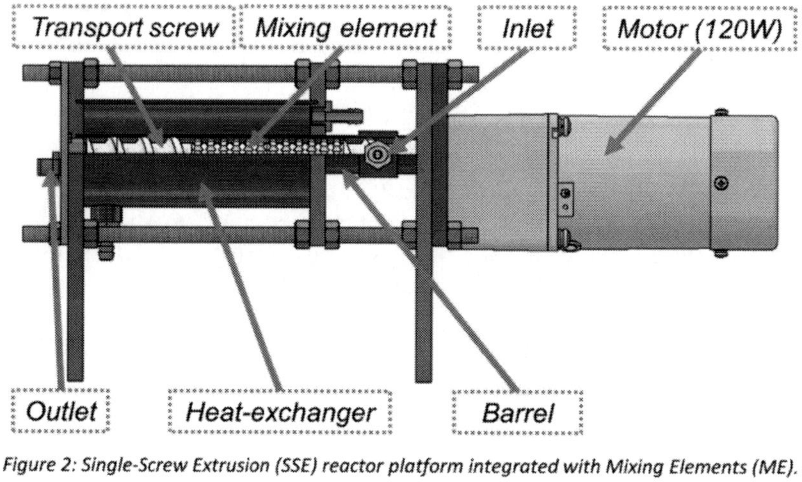

Figure 2: Single-Screw Extrusion (SSE) reactor platform integrated with Mixing Elements (ME).

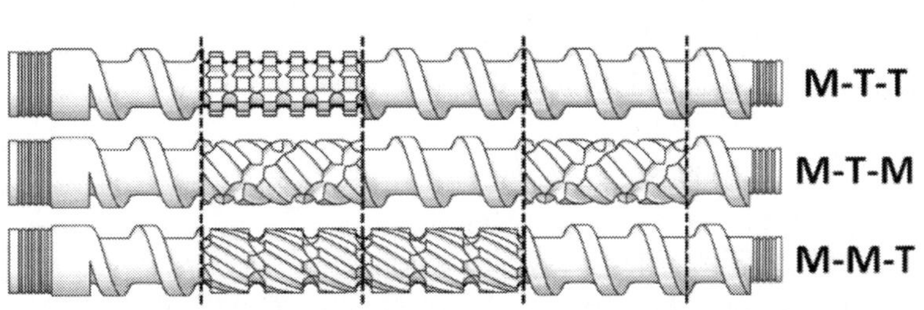

Figure 3: Different extrusion-screw ME-configurations (with T = transport element and ME = mixing element)

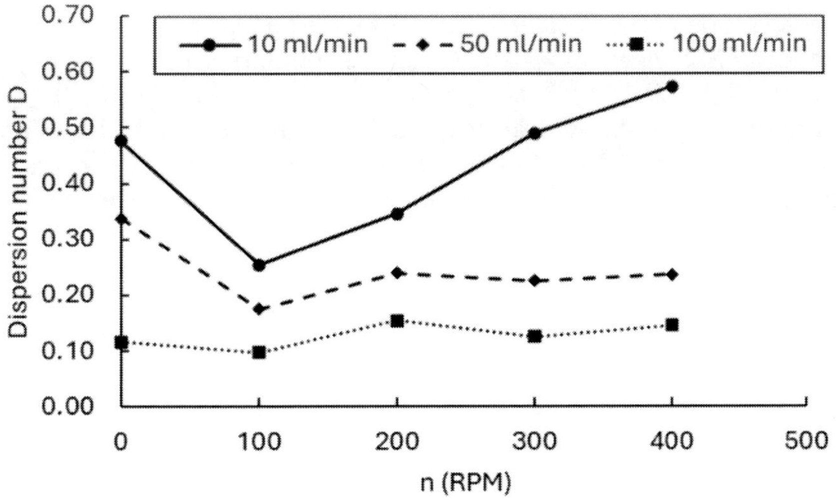

Figure 4: Effect of screw rotational speed and feed flowrate on the axial dispersion (macromixing) in the extrusion reactor. ME: Normal Pin, Configuration: M-TM, viscosity: 1mPa.s.

1c Mixing within Confined Impinging Jet Mixers: Innovations in Monitoring for Novel Applications

Cedric Devos, Yi Wei, Saikat Mukherjee, Sandip Mondal, George Barbastathis, Torsten Stelzer, Richard Braatz, Allan Myerson

Confined impinging jet mixers (CIJs) are passive mixing devices with high throughput capabilities used for rapid mixing (within ms) [1]. They consist of dual co-linear jets flowing oppositely and colliding head-on within a small confined mixing chamber [1-3]. CIJs are proven to effectively, robustly, and reproducibly intensify mixing-limited processes [3]. However, as new applications for CIJs continue to emerge, for instance in the production of complex nanoparticles (e.g., for encapsulating novel therapeutics or for manufacturing nanostructures with enhanced functionality), the demand for the development of real-time / in-situ monitoring techniques becomes increasingly important [3] to understand the processes and to address potential regulatory requirements.

In this study, we initially highlight the need for innovative monitoring technologies. Both the essential requirements as well as the intricate challenges are discussed. Then, our progress in establishing real-time / in-situ monitoring of mixing within a commercial CIJ (Knauer) is presented. The feasibility of various techniques is evaluated, including but not limited to spatially-resolved dynamic light scattering (NanoFlowSizer, Inprocess LSP) and UV/vis spectroscopy (Avantes). These techniques are used to progress to our goal of achieving controlled production of mRNA-loaded lipid nanoparticles, a process contingent upon effective mixing.

[1] Johnson, B. K.; Prud'homme, R. K. Chemical Processing and Micromixing in Confined Impinging Jets. AIChE J. 2003, 49 (9), 2264–2282. https://doi.org/10.1002/aic.690490905.

[2] Santos, R. J.; Sultan, M. A. State of the Art of Mini/Micro Jet Reactors. Chem. Eng. Technol. 2013, 36 (6), 937–949. https://doi.org/10.1002/ceat.201200678.

[3] Devos, C., Mukherjee, S., Inguva, P., Shalini S., Wei, Y., Mondal. S, Yu, H., Barbastathis, G., Stelzer, T., Braatz, R.D., Myerson, A.S. Impinging Jet Mixers: an Exhaustive Review of their Mixing Characteristics, Performance Considerations, and Applications. In preparation.

Acknowledgements: This research was supported by the U.S. Food and Drug Administration under the FDA BAA-22-00123 program, Award Number 75F40122C00200.

1d In Search of Optimal Mixing Conditions for Rheology Modifiers in Agitated Tanks

Dinesh Kumar, Svetlana Zak

Rheology modifiers are ubiquitous in the chemical, food, pharmaceuticals, construction, and cosmetic industries. They are used to enhance stability, viscosity, texture and spreadability across various industries for a wide range of formulations. The effectiveness of rheology modifiers in delivering the target rheology and desired function depends on several factors including the concentration, ionic conditions, and mixing environment in the dispersion medium in the mixing tank. In a mixing tank, the placement of impeller relative to rheology modifier location is critical for process success due to non-uniform shear rate distribution. In some cases, too low of a shear rate environment can cause incomplete activation of rheology modifier resulting in unpredictably long equilibrium times for desired function. In other cases, high shear rate environment can lead to reduced interactions between the rheology modifier and dispersion medium. This work utilizes a hybrid computational and experimental approach to investigate the changes in flow-behavior of dispersion medium with rheology modifier concentration and effectiveness of rheology modifiers in providing a desired rheology as a function of mixing parameters such as shear rate, energy dissipation rate, and impeller intensity in an agitated tank. Overall, it is demonstrated that it is critical to control the local mixing environment for achieving optimal rheology in formulations.

1e Mixing Times in Multiple Shaft, Multiple Impeller (MSMI) Stirred Tanks with Anchors in the Laminar and Transitional Regime

Darius Khoshdel, Waldo Rosales Trujillo, Mark J. H. Simmons, Andrew Ingram, Thomas Abadie

1. **Introduction**

Complex rheological fluids can develop problematic behaviours like 'dead zones' and caverning during stirred tank mixing due to microstructure rearrangement, resulting in uneven mixing. Such effects are common amongst formulated products like hair conditioner, paint and mayonnaise. Multiple shaft, multiple impeller (MSMI) stirred tanks with an anchor offer advantages for mixing these fluids due to their multifaceted approach. Many different impeller types may be installed for different purposes, operated at different heights and speeds, and used at different process stages, e.g., pumping, shearing/aggregate break-up or elimination of dead zones. The flexibility of the MSMI mixer with an anchor makes it suitable for a wide range of mixing applications. However, there is limited literature on optimising its performance.

Insights from studies on similar multiple impeller mixers can provide some guidance. Coaxial-type stirred tanks (bottom-driven anchor and top-driven agitator) shouldn't operate in counter-rotation mode as it increases power requirements without enhancing mixing [1]. For the inner impeller choice, radial impellers are preferred over axial ones because they generate more circulation loops which are disrupted by the anchor and contribute to mixing. Dual inner impeller setups, compared to single inner impeller designs, are found to reduce mixing time for a given power density, P_v, and improve efficiency [2]. In a dual-shaft setup without an anchor (e.g., Rushton-Rushton), optimal mixing involves significant height and eccentricity differences between the agitators, along with a combination of radial and axial impeller. However, counter-rotation is more effective in this arrangement [3]. The dynamics of an MSMI with an anchor are likely to incorporate heuristics from both mixer types.

2. **Methods**

The present work aims to develop the understanding of MSMI stirred tanks with anchors by visualising the mixing process via the acid-base decolourisation technique. This technique leverages pH indicator colour change to observe mixing time and dead zones via stoichiometric addition of an acid or base as a passive scalar. In this experiment, a 93% v/v glycerol solution ($\mu \approx 0.4$ Pa.s) chosen as the viscous, Newtonian fluid containing 0.3% w/w bromocresol purple was used in a 19L flat-bottomed tank ($T = 289$ mm) equipped with a top-driven anchor ($D_a = 270$ mm) and an eccentric impeller ($D_e = 60$ mm). Both Rushton and down-pumping pitched blade turbine (PBT_d) impellers were examined. Torque and power measurements were recorded using strain gauge transducers (TorqSense SGR510, Sensor Technology). Figure 1 highlights the relative simplicity of this technique.

To initiate each experiment, the glycerol solution was set to purple by adding aq. NaOH (10M). Bromocresol purple changes from purple (pH > 6.8) to yellow (pH < 5.2) when acid (aq. 10M HCl) is added. To ensure observation of macromixing rather than micromixing, an acid-to-base ratio of at least 2 was maintained upon addition. The acid was pre-mixed with a small sample of the glycerol solution to minimise viscosity/density differences between the acid and bulk solution. Videos were captured using a Fujifilm X-T3 digital camera (1080p and 50 fps) and processed using the Python image processing library OpenCV. The post-processing script identifies frames where the anchor is squarely oriented relative to the camera, applies a mask and calculates mixedness as the ratio of pixels above a threshold 'mixed' image to the total coloured pixels. A comprehensive exploration of this technique can be found in the study of Cabaret et al. (2007) [5].

3. **Results**

A 20-run, 2-level screening Design of Experiments (DoE) (resolution IV) was conducted to explore the main effects of 7 variables of the MSMI and ascertain their significance. Eccentricity ($e/R = 0.38$-0.55) and clearance ($C/T = 0.35$-0.50) of the eccentric impeller, eccentric impeller type (Rushton or PBT_d), rotation mode (co- and counter-rotation), acid addition point ($r/R = 0.35$-0.90) and anchor and eccentric impeller speeds (20-60 and 300-750 rpm, respectively) were studied. Therefore, both the laminar and early-transitional regimes were traversed ($Re_{anchor} = 75$-230 and $Re_{eccentric} = 55$-140).

The results show t95 values ranging from 66.26-600.00 s (capped due to poor mixing). Figure 2 demonstrates the spread as a function of power density, P_V. Unsurprisingly, t95 was shown to decrease with P_V but the general trend did not imply efficient mixing. Two of the fastest mixing runs, with t95 values of 77.55 s and 84.57 s, require power density inputs of 632.5 $W.m^{-3}$ and 294.2 $W.m^{-3}$, respectively, evidencing the potential for fast mixing with lower energy requirements through greater understanding of MSMI operation. The most influential factors on P_V were the anchor and eccentric impeller speeds which, in turn, appeared as the only statistically significant factors influencing t95, i.e., $p\text{-value} < 0.05$. However, given the nature of 2-level, resolution IV DoE, it is assumed that the relationship between impeller speed and t95 is linear and interaction factors (e.g., anchor speed*eccentric impeller speed) are unimportant since two-factor interactions are aliased with each other. It is highly unlikely that this is true and is one of the pitfalls to this type of experimentation.

Additionally, the results echoed those of coaxial stirred tank literature such that the counter-rotation mode increased anchor power demand [1]. In this mode, the two impellers appear to be working against each other which results in an increased demand for the anchor only (N.B., anchor speed had no significant effect on eccentric impeller power). Rotation mode was not deemed a statistically significant factor for predicting t95 so it is likely that the co-rotation mode is optimal for MSMI mixers as well. The two other statistically significant factors relating to power demand were impeller type and eccentricity. Of course, the Rushton impeller had greater power consumption compared to the PBT_d which agrees with power curve data [5], however, the eccentricity of the Rushton or PBT_d was seen to decrease anchor power requirements and further investigation is required to understand the mechanics behind this effect.

Neither eccentric impeller clearance nor injection point were found to be statistically significant relative to t95 or power consumption and will be dropped from further study. This is not to say that they have no effect but compared to the other 5 variables (eccentricity, eccentric impeller type, rotation mode and anchor and eccentric impeller speeds) their impact is insignificant. Since it is likely that non-linear relationships and interaction effects are present, a higher resolution DoE is necessary so it is beneficial to reduce the number of variables studied to minimise the total number of runs required for future study.

4. Conclusions

The present work lays the foundation for further, more detailed study of MSMI stirred tanks with anchors in the laminar and early-transitional regime for viscous, Newtonian fluids. Mixing time, t95, and individual impeller power requirements were observed and it was seen that eccentricity, eccentric impeller type, rotation mode and anchor and eccentric impeller speeds were statistically significant contributors to MSMI performance. However, the screening-type DoE carried out in this work assumes a linear relationship between these variables and outputs and disregards any potential interaction effects which is unlikely. Further work, using a higher resolution DoE (resolution V or greater), is required to determine the truth of the above statement. It is interesting to note the similarity between MSMI and coaxial mixers, i.e., co-rotation mode reduces power demand with no t95 penalty, but further work is required to understand other effects such as the impact of impeller eccentricity on the anchor power draw (where no inference can be drawn from coaxial literature). 2 of the 7 studied variables, eccentric impeller clearance and injection point, were found to be insignificant factors in terms of t95 and power draw and will be dropped from future study. Therefore, a reduced number of variables will be taken forward in the future work which sets out to explore MSMI operation to greater degree of detail, accounting for non-linear relationships and interaction effects, and study the impacts of more complex, non-Newtonian fluid rheology on its operation and performance.

References

[1] S. Foucault, G. Ascanio, and P. A. Tanguy, "Mixing Times in Coaxial Mixers with Newtonian and Non-Newtonian Fluids," *Industrial & Engineering Chemistry Research*, vol. 45, no. 1, pp. 352–359, Dec. 2005. doi:10.1021/ie048761o

[2] A. Kazemzadeh, F. Ein-Mozaffari, A. Lohi and L. Pakzad, "A new perspective in the evaluation of the mixing of biopolymer solutions with different coaxial mixers comprising of two dispersing impellers and a wall scraping anchor", *Chemical Engineering Research and Design*, vol. 114, pp. 202–219, Oct. 2016. doi:10.1016/j.cherd.2016.08.017

[3] S. Wang et al., "CFD simulation of hydrodynamics and mixing performance in dual shaft eccentric mixers", *Chinese Journal of Chemical Engineering*, vol. 62, pp. 297–309, Oct. 2023. doi:10.1016/j.cjche.2023.03.004

[4] F. Cabaret, S. Bonnot, L. Fradette and P. A. Tanguy, "Mixing Time Analysis Using Colorimetric Methods and Image Processing", *Industrial & Engineering Chemistry Research*, vol. 46, no. 14, pp. 5032–5042, Jun. 2007. doi:10.1021/ie0613265

[5] R. L. Bates, P. L. Fondy and R. R. Corpstein, "Examination of Some Geometric Parameters of Impeller Power", *Industrial & Engineering Chemistry Process Design and Development*, vol. 2, no. 4, pp. 310–314, Oct. 1963. doi:10.1021/i260008a011

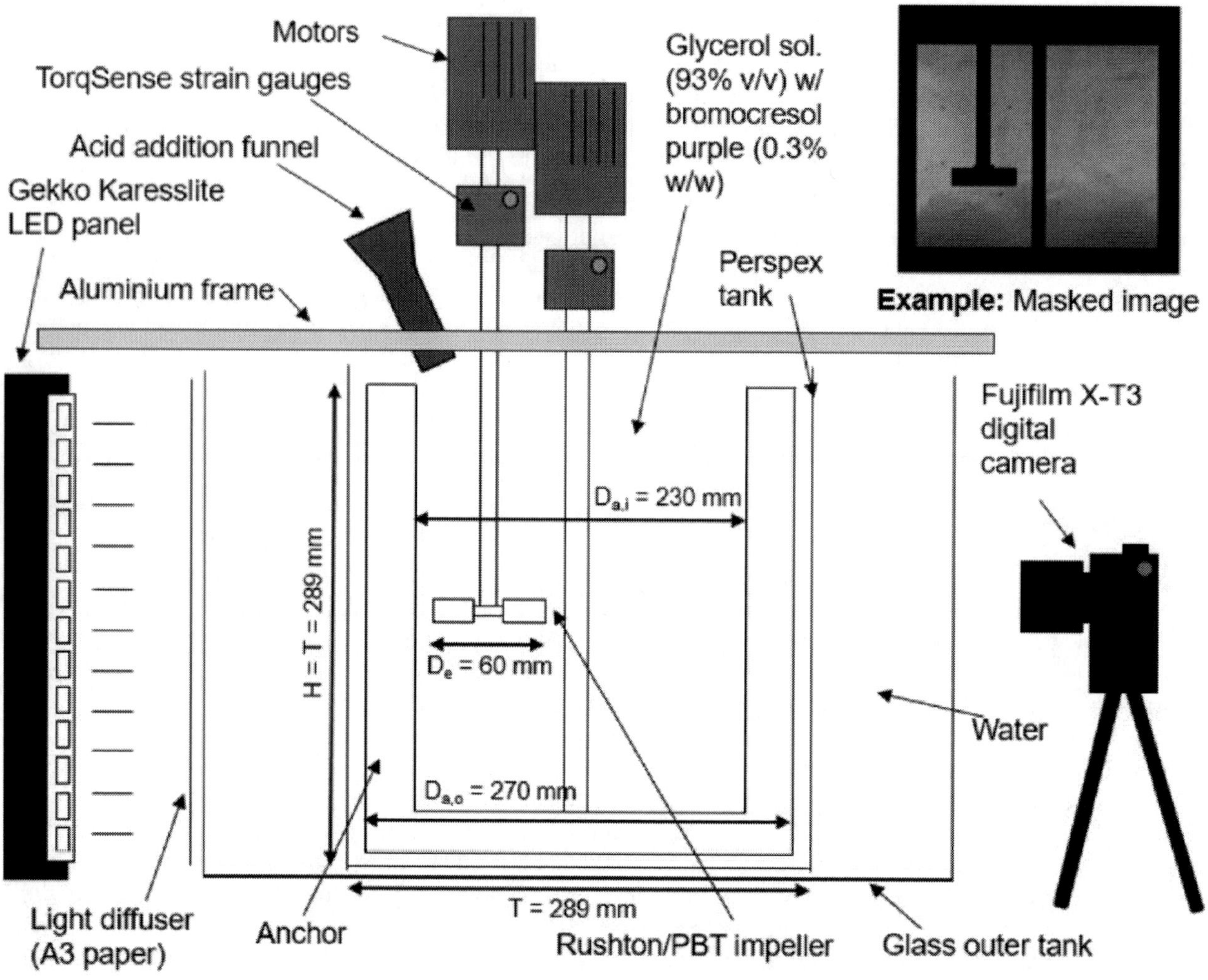

Figure 1. Schematic diagram of the acid-base decolourisation set-up for the MSMI stirred tank.

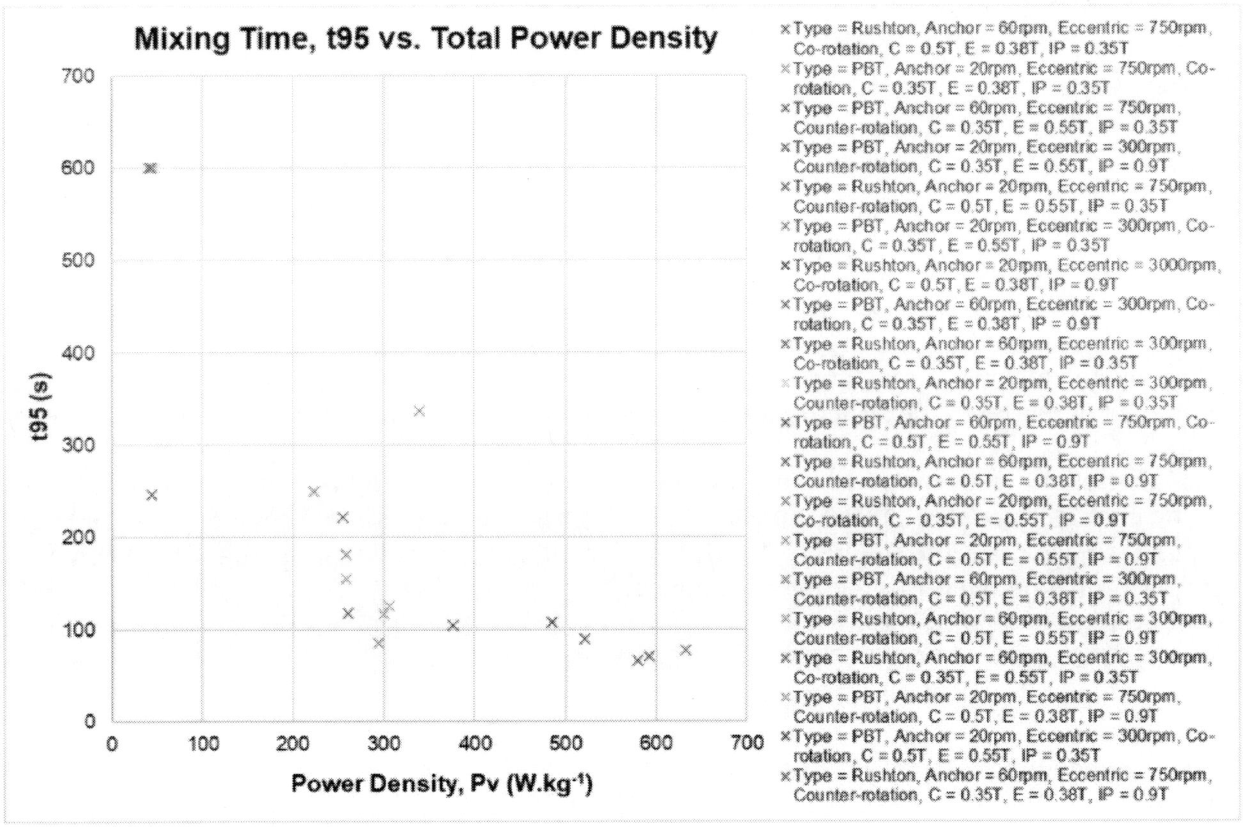

Figure 2. Mixing time, t95, as a function of power density, P_v for all 20 runs of the resolution IV screening DoE.

1f Improvement in Understanding of the Sodium Sulfite Steady-State Method for Gas-Liquid Mass Transfer Measurement in Precision Fermentation Applications

Cormac Ryan, Sarah Lanzafame, Kevin Logsdon, Casey Stein, Jason Giacomelli, Richard Grenville

One of the recognized methods for mass transfer measurements in Gas-Liquid systems is the excess sodium sulfite method catalyzed by cobalt. In this method, the stoichiometric sodium sulfite required to consume the dissolved oxygen from a system is calculated and a specific quantity larger than this is added to the gas-liquid system. The time taken to consume this known quantity of excess sulfite can then be equated to a mass transfer rate of oxygen into the system. This measured mass transfer must be corrected to represent the system of interest for the following reasons:

- Addition of ionic components is known to influence the mass transfer rate (1). This effect must be corrected for as additional sulfite is added. Dissolved ions also affect the interfacial tension and the gas molecule solubility. This also limits the number of experiments which can be performed before replacing the bulk liquid.
- Variation of temperature in the gas-liquid system is also known to influence the mass transfer rate. There is an ASCE (ASCE, American Society of Civil Engineers, 2022) correction factor for this for the standard sulfite method (2).

A full understanding of the effect of these variables is key to:

- Accurately representing the baseline mass transfer rate and drawing the correct conclusions from the results of the experiments.
- Maximizing the number of trials per volume of bulk liquid while ensuring that accuracy is maintained.

As a solution provider and researcher for Gas-Liquid mixing processes, SPX FLOW has an interest in understanding these effects further. While there are other methods that do not suffer the same drawbacks of the sulfite method, the sulfite method remains a very cost effective and quick technique for assessing the mass transfer potential of a system. The effects of temperature and total dissolved solids or ionic strength can be accounted for and thus eliminated from the data. An analysis of recent mass transfer measurement experiments was conducted to investigate the effects of these variables and improve the sulfite technique for measuring mass transfer in gas-liquid stirred tank reactor systems.

The testing was performed at a single scale (T = 0.46m, H/T = 2.5) with technical grade sodium sulfite over a period of weeks. The water temperature was not controlled and varied between 10 – 17°C. Regression models were fit using R statistical programming language and different regression models were compared using Leave One Out Cross Validation (LOOCV) to prevent overfitting.

It was concluded that the effect of temperature is not fully represented by the ASCE correction factor. By including temperature as a power law variable, the model RMSPE (root mean square percentage error) was reduced from 5.9% to 3.5% (Approx 40% reduction). This reduction in error led to more robust interpretation of the impact of other variables in precision fermentation processes.

It was also concluded that the effect of sulfite concentration varied when measured on different dates for different physical setups. This effect could not be explained by the additional temperature correction factor. A custom polynomial fit for sulfite correction was used for some sets of trials due to this variation. It was noted that using a power-law variable to represent this in regression resulted in only a minor effect on accuracy.

References:

1. Van't Riet, "Review of Measuring Methods and Results in Nonviscous Gas-Liquid Mass Transfer in Stirred Vessels" Industrial & Engineering Chemistry Process Design and Development, vol. 18, 1979.
2. ASCE, American Society of Civil Engineers. (2022). *Measurement of Oxygen Transfer in Clean Water.* Reston, Virginia: ASCE/EWRI.

546a Mass Transfer in a Von Kármán Bioreactor to Improve Fermentation

Esther Jokodola, Daniel Bun, Krishnamurthy Ravichandar, Zengyi Shao, R. Dennis Vigil

Heterogeneity in bioreactors poses a challenge for commercialization of many biochemical processes as it can significantly impact product quality and process scaleup. To tackle these challenges, novel approaches are needed to better manage environmental trajectories experienced by microorganisms. In this work we discuss the usage of von Kármán swirling flow to both reduce reactor heterogeneity and mass transport limitations. Here, we describe the development of a benchtop von Kármán reactor (vKR) and compare its performance with an analogous stirred tank bioreactor in a case study for producing citramalate in a yeast strain. Citramalate is the precursor of poly methyl methacrylate, a transparent material known as acrylic glass or plexiglass.

Experiments were performed in a 3-L vKR to characterize the volumetric gas-liquid mass transport coefficient (k_La) as well as gas holdup as a function of impeller speed and gas flow rate, from which dimensionless correlations were developed. Compared with stirred tank reactors of similar volume, the values of k_La and gas holdup produced in the vKR are significantly larger. When compared with analogous fermentation experiments performed in a conventional stirred bioreactor, it is evident that even small volume (2-L) stirred bioreactors can be susceptible to mass transfer limitations, which in turn limits their use to quickly grow biomass and in characterizing biokinetics. In such situations, a von Kármán bioreactor may provide an alternative to overcome this problem. Lastly, we examine the possibility of further improving the von Kármán bioreactor efficiency by optimizing oxygen feed rate to manipulate metabolic flux distributed between organism growth and production of the target product.

546b Flash Nanoprecipitation As an Agrochemical Nanocarrier Formulation Platform: Phloem Uptake and Translocation after Foliar Administration

Kurt Ristroph, Yilin Zhang, Valeria Nava, Jonas Wielinski, Hagay Kohay, Andrew M. Kiss, Juergen Thieme, Gregory V. Lowry

The increasing severity of pathogenic and environmental stressors that negatively affect plant health has led to interest in developing next-generation agrochemical delivery systems capable of precisely trafficking active agents to specific sites within plants. Many agrochemicals of interest stand to benefit from formulation into colloidal suspensions for improved wettability, flowability, and possibly uptake into plants, but the techniques used to prepare such formulations (cost, scale, etc.) must be carefully considered. In this work, we adapt Flash NanoPrecipitation (FNP), a scalable antisolvent precipitation technology driven by rapid turbulent mixing, to prepare organic core-shell nanocarriers to evaluate for uptake and translocation in tomato plants following foliar or root administration.

NCs ranging in diameter from 55nm to 200nm, with surface zeta potentials from -40mV to +40mV, and with seven different shell material properties were prepared and studied. Shell materials included synthetic polymers poly(acrylic acid), poly(ethylene glycol), and poly(2-(dimethylamino)ethyl methacrylate), naturally-occurring compounds fish gelatin and soybean lecithin, and semisynthetic hydroxypropyl methylcellulose acetate succinate (HPMCAS). NC cores contained a gadolinium tracer for tracking by mass spectrometry, a fluorescent dye for tracking by confocal microscopy, and model hydrophobic compounds (alpha tocopherol acetate, polystyrene) that could be replaced by agrochemical payloads in subsequent applications. After foliar application onto tomato plants with Silwet L-77 surfactant, internalization efficiencies up to 85% and NC translocation efficiencies up to 32% were observed. Significant NC trafficking to stem and roots suggests a high degree of phloem loading for these materials. Results were corroborated by confocal microscopy and synchrotron X-ray fluorescence mapping. NCs stabilized by cellulosic HPMCAS exhibited the highest degree of translocation, followed by formulations with significant surface charge. The results from this work indicate that an industrially-viable pharmaceutical nanoformulation process (FNP) using biocompatible materials like HPMCAS are promising agrochemical delivery vehicles, and shed light on the optimal properties of organic NCs for efficient foliar uptake, translocation, and delivery.

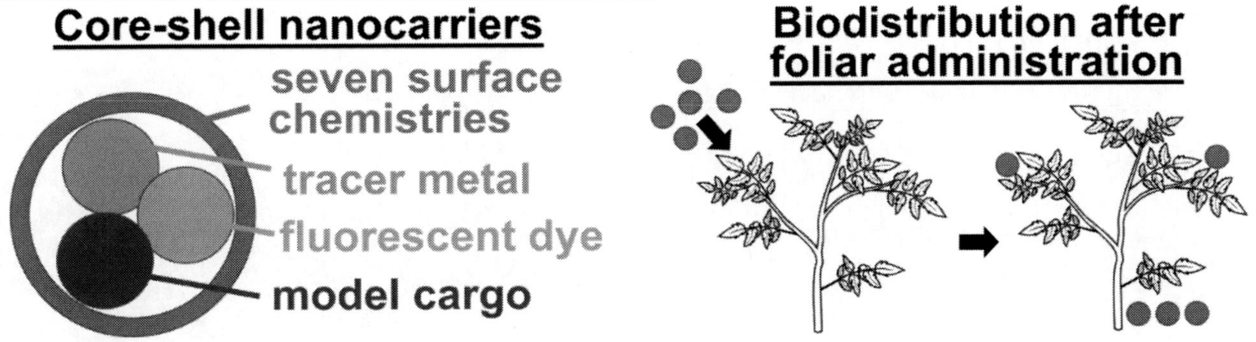

546c Multiphase Simulation of Bioreactors to Improve Biofuel Production at Scale

Mohammad Rahimi, Julia Ream, Nicholas Wimer, Hari Sitaraman, Marc Day, Davinia Salvachua

There is an increasing demand for commercial aviation fuels with low carbon footprint, such as sustainable aviation fuels (SAF), due to the global air transport industry's commitment to carbon-neutral growth and net-zero carbon emissions by 2050 [1]. The main challenge is to increase the production of SAF to meet increasing world-wide demand at costs that are comparable to those of existing fossil fuels. At small scales in a laboratory setting, gas fermentation bioreactors have been demonstrated to generate exceptionally high product yield. However, when scaled up for mass industrial production, these systems exhibit significant performance reductions. In this work, we use computational models to elucidate the key factors affecting the efficiency of bioreactors at industrial scale.

A Michaelis–Menten kinetic model for the growth and uptake rates of the relevant microorganisms, substrate and other dissolved species, is developed using data from lab-scale stirred tank bioreactors. This model is coupled with a multiphase fluid dynamics model [2,3] and validated using data from the lab-scale bioreactor, then used to estimate the performance of an industrial-scale stirred tank bioreactor. Detailed analysis of the hydrodynamics as well as the substrate concentrations inside this bioreactor enables us to identify the key factors that impact production efficiencies at industrial scale.

References:

[1] https://www.iata.org/en/programs/environment/sustainable-aviation-fuels/

[2] Rahimi, M. J., Sitaraman, H., Humbird, D., and Stickel, J. J. (2018). Computational fluid dynamics study of full-scale aerobic bioreactors: Evaluation of gas–liquid mass transfer, oxygen uptake, and dynamic oxygen distribution. *Chemical Engineering Research and Design, 139*, 283–295

[3] Sitaraman H, Lischeske J, Stickel J & Lu Y (2023). A reacting multiphase computational flow model for 2,3-butanediol synthesis in industrial-scale bioreactors, *Chemical Engineering Research and Design,* 197, 2023, 38-52

546d Analysis of Mass Transfer in a Three-Phase Taylor-Couette Reactors: Implications for Mixing in Fermentation

Krishnamurthy Ravichandar, Daniel Bun, Zengyi Shao, R. Dennis Vigil

Designing novel bioreactors has recently garnered significant attention as a cutting-edge topic in biomanufacturing. The utilization of Taylor vortex flow, generated in the annular region between a rotating inner cylinder and a stationary concentric outer cylinder, has attracted increasing interest for its application in multiphase reaction and separation processes. In our study, we focus on applying Taylor vortex flow to achieve simultaneous fermentation and in situ product recovery, which alleviates product inhibition on growth, enhance productivity, and increase genetic stability of the culture. To effectively implement this concept, a deeper understanding of mass transfer between three phases (including two immiscible liquids and gas phase) is crucial. In this study, we designed experiments using a semi-batch liquid-liquid-gas Taylor vortex device to ferment a fatty-alcohol producing strain with a continuous gas feed to investigate volumetric mass transfer coefficients (k_La). Fatty alcohols are commonly used as emollients, thickeners, and surfactants in various industries. Dodecane is often added into the fermentation medium to simultaneously extract the relatively toxic fatty alcohols from the broth. The results indicate that, for gas-liquid mass transfer, k_La increases with the rotation speed of the inner cylinder (corresponding to azimuthal Reynolds number). However, beyond a critical value, further increases in rotation speed have minimal impacts. Moreover, increases in gas flow rate in the range of 60 and 250ml/min significantly enhance k_La at fixed azimuthal Reynolds numbers. In contrast, for liquid-liquid mass transfer between the aqueous phase (fermentation medium) and the organic phase (dodecane), higher azimuthal Reynolds numbers in the range of 13611 and 27222 were found to enhance interphase mass transfer at fixed gas flow rates. Additionally, increasing the gas flow rate was observed to enhance the liquid-liquid mixing rate, resulting in higher k_La between the aqueous and organic phases. In the context of bioreactors, oxygen mass transfer from air to fermentation broth plays a critical role in balancing biomass accumulation and metabolic functions closely relevant to product synthesis. By optimizing k_La through manipulation of gas flow rate and cylinder rotation speed in a Taylor vortex bioreactor, we are able to improve the availability of oxygen in the fermentation media, balance metabolic activities and therefore enhance overall productivity.

Improvement of mixing characteristics of an atypical unbaffled vessel with unsteady rotation speed impeller

Authors: Jan Haidl[a], Ondřej Gebouský[a,b], Kristýna Idžakovičová[a,b], Martin Pivokonský[a], Zbyněk Kálal[c]

Keywords: Chemical Engineering Practice-Fundamentals-Fluid Mechanics, Chemical Engineering Practice-Mixing

[a] Institute of Hydrodynamics of The Czech Academy of Sciences, Pod Paťankou 30/5, Prague 6, Czech Republic

[b] University of Chemistry and Technology Prague, Technická 5, Prague 6, Czech Republic

[c] MSD Czech Republic s.r.o., Svornosti 3321/2, Prague 5, Czech Republic

Motivation

Mixing of liquids is a key unit operation across diverse industrial processes, frequently employing standard mixed vessels with four wall baffles and a central impeller. However, in some industries, such as biotechnology and pharmaceuticals, this mixing-efficient configuration is not preferred because the presence of built-ins or shaft seals contradicts the increased requirements on equipment cleanability and sterility. A common alternative solution that addresses these requirements is an unbaffled stirred bioreactor with an off-center, magnetically driven impeller [1]. The eccentrical agitator position mitigates the central vortex formation. Nevertheless, since magnetic propulsion requires the impellers to be placed at the vessel bottom, the maximum agitator size is limited due to the vessel bottom curvature. This vessel configuration has not been extensively studied, which results in a lack of fundamental understanding and characterization. Moreover, while the standard baffled vessels ensure rapid batch homogenization, these unconventional stirred bioreactors are prone to the development of a poorly mixed segregated region, prolonging homogenization times, particularly with higher liquid levels [2]. The segregated zone may pose risks to microorganisms, especially during aerobic processes. Our objective is to enhance mixing while preserving the benefits of unbaffled setups with magnetically driven impellers.

Introduction

The literature survey yielded a promising approach leveraging variation in agitator rotational speed or direction - dynamic mixing protocol (DMP) - implemented by Yao et al. to destabilize segregated regions and thereby enhance laminar mixing of viscous liquids [3]. The application of forward-reverse mixing to enhance turbulent flow in unbaffled vessels was investigated in pioneer studies by [4, 5]. Although they did not report a comparison of mixing times in the unbaffled vessels, they observed suppression of tangential liquid rotation and central vortex formation. Studies also address an increase in the power number to values similar to or slightly higher than those in the fully baffled configuration, which indicates an increase in peak shear stresses and the risk of batch degradation. We assume that this increase in unbaffled vessels with a central impeller is associated with the transition of the flow to the turbulent regime, while in the configuration of interest, the stirrer already operates in the turbulent region at a constant speed mixing [2].

We present a novel and comprehensive study combining experimental and validated CFD data to demonstrate the DMP potential to enhance mixing in unbaffled vessels with magnetically driven impellers. The study focuses on the effect of DMP on peak shear stresses that directly affect cultured microorganisms or shear-sensitive products. We analyzed the impact of DMP on segregated region destabilization and overall mixing time in a pilot-scale, unbaffled bioreactor with an off-centered impeller. The CFD methodology used in this work was validated using experimental data on mixing power and mixing times. Thus, it can serve as a standalone framework for further study and design of these systems.

Experimental setup and methodology

The experiments were carried out in a cylindrical pilot-plant vessel with a diameter of T=390 mm, using tap water at two different liquid levels (H=440 mm and 600 mm). The tank was operated in an unbaffled arrangement with low clearance (C=10 mm) and an eccentrically mounted impeller, mimicking a bioreactor with a magnetically driven impeller. Two Rushton turbines with D=60 mm and 100 mm diameters were used. The shaft torque was measured using a *Burster 8645* torque meter (±2.5 Nm, accuracy <1% F.S.). The experimental setup is depicted in Fig. 1.

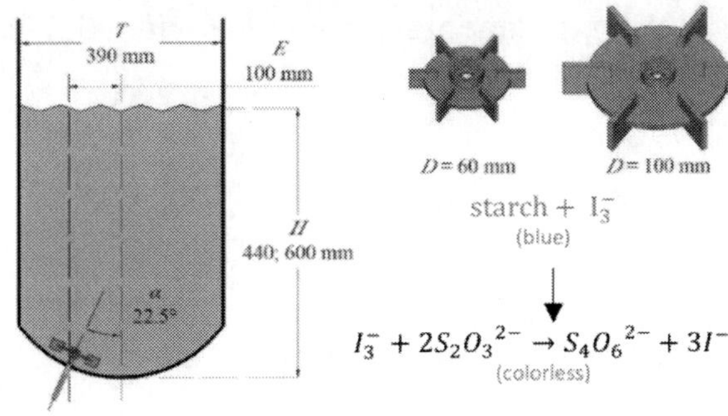

Fig. 1: Sketch of the experimental tank, two Rushton turbine impellers, and the equation of the decolorization reaction.

The primary rotational speeds (N_P) corresponding to power inputs P_P of 10 W/m^3 and 100 W/m^3 were used for both impellers and liquid levels. DMP with rotational speeds alternating between N_P and various secondary speeds N_S was employed. The protocol consists of phases of constant speeds N_P and N_S with a linear acceleration/deceleration between the phases. The duration of the constant-speed phase (5-20s) was optimized for any N_P/N_S combination; the acceleration/deceleration phase lasted 3 s in the cases of constant rotation direction and 6 s in the cases of changing rotation direction. The mixing characteristics were measured using the decolorization method based on the iodine-starch-sodium thiosulfate system measured the mixing characteristics. To measure the standard mixing time θ, the color tracer was injected near the impeller. To evaluate the segregated zone stability, the tracer was injected directly into the segregated zone, and the time τ required to vortex decolorization was measured. Each measurement was performed at least three times under given conditions.

Numerical flow and tracer transport simulations were performed using a modified transient flow solver in the OpenFOAM© environment. The computational domain matched the true geometry of the pilot-plant vessel filled up to H=600 mm. The mesh consisted mostly of hexahedral elements, locally refined close to the impeller, walls, and inside the MRF region. The k-ω SST DES model for turbulence modeling was employed. The impeller rotation was modeled using the MRF approach. First, a converged steady-state flow simulation was obtained using k-ω SST turbulence model, which was used as an initial condition for the following 60-s transient flow simulation with the impeller rotating according to the desired

DMP protocol. Then, the initial concentration of a tracer was prescribed to a small volume inside the vessel positioned either close to the impeller or inside the vortex near the water surface, and the simulation continued for an additional 60 seconds. The tracer dispersion dynamics was evaluated as the development of the coefficient of variance (CoV) of the tracer concentration.

Results and discussion

To investigate the impact of DMP on impeller power characteristics, the torque *M* was measured for a 100mm impeller operating in the following ways: (i) a constant speed of $5.9\ s^{-1}$, (ii) periodically stopping the impeller and (iii) periodically reversing impeller. Concurrently, numerical simulations of these three cases were also evaluated. When operating at a constant speed, the specific power input *P/V* remains at 100±8 W/m³. With the DMP, neither stopping the impeller nor reversing the impeller caused a significant increase in the peak torque compared to the continuously running impeller. The results of the experiments and simulations are illustrated in Fig. 2. Evidently, employing the DMP does not substantially elevate the impeller power number and peak shear stress. The results also show that the numerical simulation captured the experimental power characteristics well.

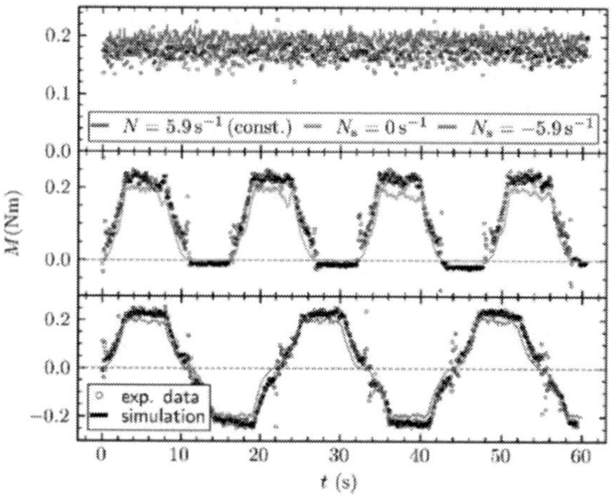

Fig. 2: Experimental and numerical simulation data of the evolution of shaft torque under constant speed mixing and dynamic mixing protocol conditions.

Contrary to the small effect of DMP on impeller power characteristics, its employment has a significant impact on the segregated zone stability. As illustrated in Fig. 3, the segregated zone is most stable in the case of constant mixing (r_P =1). However, even a minor change in rotation speed results in a relatively sharp decrease in the time τ required to decolorize the segregated zone. The best results were achieved with the rotation direction fully reversed (r_P =-1), resulting in the prevention of central vortex formation. Under these conditions, the decolorization times almost matched the mixing time.

As for the standard mixing times θ, the DMP impact differs depending on the liquid level height. At *H*=440 mm, θ remains almost independent of the DMP used. The small observed increase in θ with the impeller periodically stopping corresponds to a 50% lower total dissipated power.

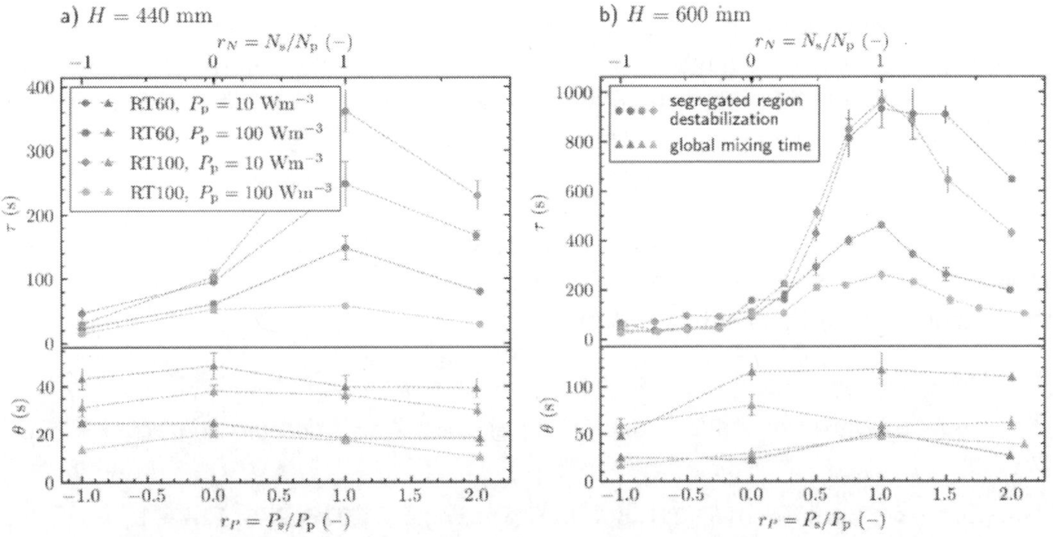

Fig. 3: Impact of the dynamic mixing protocol on the destabilization of segregated region τ and overall mixing time θ in unbaffled bioreactor with magnetically driven impeller operated with liquid levels of (a) *H*=440 mm and (b) *H*=600 mm.

At *H*=600 mm, disrupting the segregated zone stability using DMP also results in a noticeable decrease in θ in most cases. In the cases where a slight increase in θ is shown, the cause lies in the different speed change frequencies that are optimal for either segregated zone destabilization or entire vessel homogenization. As experimentally proven, it is possible to find the speed change frequency at which θ and τ are both reduced significantly.

Tracer spreading was also simulated using CFD for the same cases as for the torque evolution (i-iii); the results are shown in Fig. 4. Apparently, the experimentally determined values of θ agree very well with the simulation time required for CoV to decrease below 0.05 (which, in the literature, is usually identified as the standard mixing time θ_{95}). Simultaneously, the simulated trends in the τ evolution correspond to those measured.

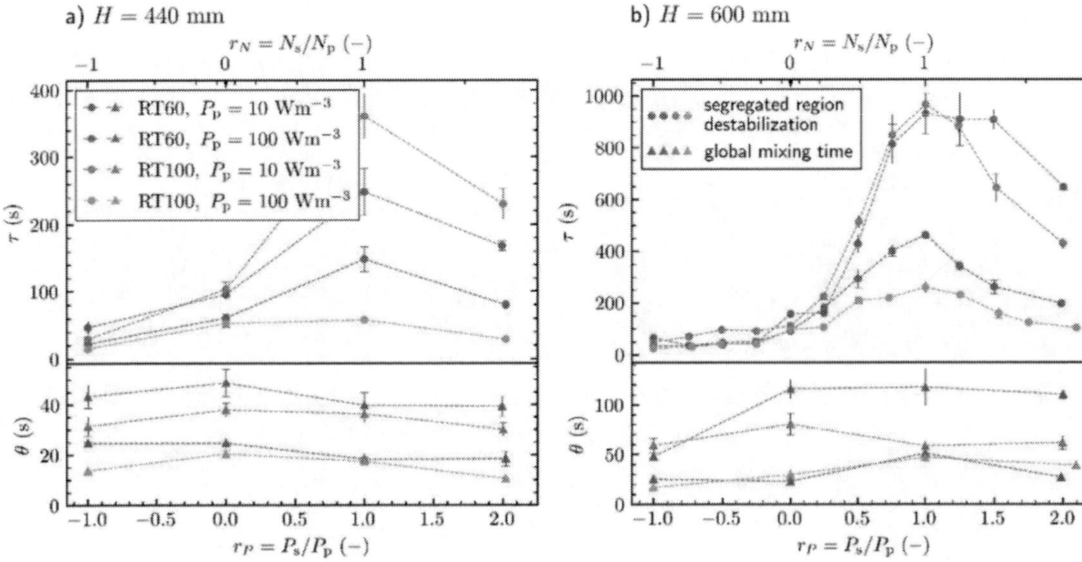

Fig. 4: Comparison of numerical simulation and experimental data on decolorization times θ and τ under the constant speed mixing and DMP conditions.

Conclusions and implications

This study presents the benefits of the dynamic mixing protocol for the mixing characteristics of a bioreactor stirred with a magnetically driven impeller. The experimental and numerical results indicate that DMP breaks down the segregated vortex zone and significantly facilitates its blending with the liquid bulk without elevated shear stress imposed by the impeller. Moreover, a lower total energy consumption is required compared to a constant-speed operation. In most cases, the DMP does not increase the measured standard mixing time and, conversely, improves it at higher liquid levels. The results, coupled with the relative simplicity of DMP, imply the potential of DMP to be used in a wide range of mixing processes across the biotechnology and pharmaceutical industries.

References

[1] Schirmer, C.; Maschke, R. W.; Pörtner, R.; Eibl, D.; *Appl.Microbiol.Biotechnol.* **2021**, 105, 2225–2242.

[2] Gebouský, O.; Idžakovičová, K.; Haidl, J.; *Chem.Eng.Sci.* **2023**, 276, 118801.

[3] Yao, W. G.; Sato, H.; Takashi, K.; Koyama, K.; *Chem.Eng.Sci.* **1998**, 53, 3031–3040.

[4] Woziwodski, S.; *Chem.Eng.Technol.* **2011**, 34, 767–774.

[5] Yoshida, M.; Shigeyama, M.; Hiura, T.; Yamagiwa, K.; Ohkawa, A.; Tezura, S.; Asia-*Pac.J.Chem.Eng.* **2007**, 2, 659–664.

546f Understanding Risk of Dissolved Gas Gradients in Sparged Systems

Matthew Flamm, Zbynek Kalal

Bioreactors, fermenters, and hydrogenation reactors involve mass transfer of a gaseous species to or from the liquid phase using sparged bubbles, and the dissolved gas is consumed or produced in a (bio)chemical reaction. The aqueous concentration of these species can determine the quality and yield of the final product and can be monitored by in-process or offline process analytical technology (PAT) or probes. An example is dissolved oxygen (DO) in a bioreactor or fermenter, where changes in DO can drive metabolic changes in the cellular metabolism. However, during scale-up it is possible to observe heterogeneity in aqueous concentrations due to mixing challenges. There is little exploration in the literature of how to estimate when such systems are operating in regimes of heterogeneity without direct measurement. This situation is common in industrial systems where often only a single probe is used to provide information about the dissolved gas concentration level for feedback control. Complex Computational Fluid Dynamics (CFD) models that incorporate bubble-liquid drag coupling, bubble breakage/coalescence, and mass transfer between the gaseous and liquid phases could provide such a prediction. CFD-based approaches are numerically expensive, but more importantly often require tuning multiple sub-model choices and/or parameters such as the initial bubble size from the sparger and the drag model form, diminishing their practical utility.

Here, the Dissolved Gas Gradient, DGG, number is developed to describe the degree of heterogeneity of dissolved gas species. The DGG number is derived from a dimensionless analysis of the various time scales, which results in a modification to the well-known Damköhler number by incorporating a description of the heterogeneity of the mass transfer of the gas species. The unknowns in the complex behavior of the gas bubbles can be lumped into a single parameter, which can be tuned if multiple measurements are available, and if measurements are unavailable, literature evidence can be used to give estimates. A greatly simplified CFD model employing the one lumped parameter is developed. This analysis shows that the DGG number alone sufficiently describes the degree of heterogeneity of the concentration of the gaseous species in the liquid phase. The importance of this analysis will be demonstrated by determining the DGG number for various cell culture operating conditions including bioreactor scale, P/V, and cell culture type.

Contents lists available at ScienceDirect

Journal of Colloid And Interface Science

journal homepage: www.elsevier.com/locate/jcis

Regular Article

Mechanistic insights into how mixing factors govern polyelectrolyte-surfactant complexation in RNA lipid nanoparticle formulation

Sophia R. Dasaro [a], Abhishek Singh [b], Pavlos Vlachos [b], Kurt D. Ristroph [a,*]

[a] Department of Agricultural and Biological Engineering, Purdue University, 225 S. University St., West Lafayette, IN 47907, USA
[b] Department of Mechanical Engineering, Purdue University, 585 Purdue Mall, West Lafayette, IN 47907, USA

GRAPHICAL ABSTRACT

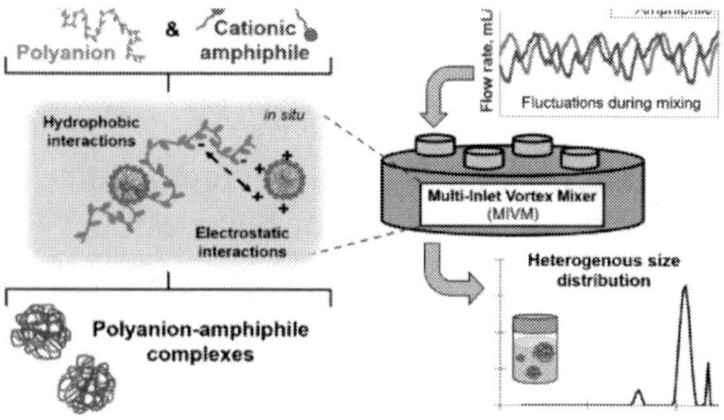

ARTICLE INFO

Keywords:
Mixing
Lipid nanoparticles
RNA
Flash NanoPrecipitation
Polyelectrolyte-surfactant complexation

ABSTRACT

Hypothesis: Lipid nanoparticle self-assembly is a complex process that relies on ion pairing between nucleic acids and hydrophobic cationic lipid counterions for encapsulation. The chemical factors influencing this process, such as formulation composition, have been the focus of recent research. However, the physical factors, particularly the mixing protocol, which directly modulates these chemical factors, have yet to be mechanistically examined using a reproducible mixing platform comparable to the industry standard. We here utilize Flash NanoPrecipitation (FNP), a scalable rapid mixing platform, to isolate and systematically investigate how mixing factors influence this complexation step, first by using a model polyelectrolyte-surfactant system and then generalizing to a typical RNA lipid nanoparticle formulation.

Experiments: Aqueous polystyrene sulfonate (PSS) and cetrimonium bromide (CTAB) solutions are rapidly homogenized using reproducible FNP mixing and controlled flow rates at different stoichiometric ratios and total solids concentrations to form polyelectrolyte-surfactant complexes (PESCs). Then, key mixing factors such as total flow rate, inlet stream relative volumetric flow rate, and magnitude of flow fluctuation are studied using both this PESC system and an RNA lipid nanoparticle formulation.

Findings: Fluctuations in flow as low as $\pm$ 5 % of the total flow rate are found to severely compromise PESC formation. This result is replicated in the RNA lipid nanoparticle system, which exhibited significant differences

* Corresponding author.
E-mail address: ristroph@purdue.edu (K.D. Ristroph).

https://doi.org/10.1016/j.jcis.2024.08.150
Received 22 April 2024; Received in revised form 4 August 2024; Accepted 19 August 2024
Available online 22 August 2024

in size (132.7 nm vs. 75.6 nm) and RNA encapsulation efficiency (34.0 % vs. 82.8 %) under fluctuating vs. steady flow. We explain these results in light of the chemical variables isolated and studied; slow or nonuniform mixing generates localized concentration gradients that disrupt the balance between the hydrophobic and electrostatic forces that drive complex formation. These experiments contribute to our understanding of the complexation stage of lipid nanoparticle formation and provide practical insights into the importance of developing controlled mixing protocols in industry.

1. Introduction

Encapsulation of nucleic acids into nanocarriers has provided a novel pathway for delivering life-saving vaccines and targeted gene therapies [1,2]. RNA lipid nanoparticles (LNPs) are a major area of this research due to the success of commercial formulations such as Alnylam's Onpattro (Patisiran), Moderna's mRNA-1273 (Spikevax), and Pfizer-BioNTech's BNT162b2 (Comirnaty) [3]. These formulations utilize electrostatic complexation between water-soluble anionic RNA and cationic ionizable lipids containing hydrophobic groups to drive precipitation and nanoencapsulation of the nucleic acid [2,4,5]. This complexation step in LNP assembly is one part of the larger nanoparticle formulation process that relies on hydrophobic interactions among the various lipids, supersaturation of these lipids, precipitation from ethanol into water, pH, ionic strength, and steric interactions between PEG-lipid-stabilized LNPs, among other factors. The many variables affecting these steps quickly become convoluted and are difficult to examine individually. Therefore, studying the complexation event in isolation from the other steps in the LNP formulation process may offer mechanistic insights generalizable to the full system.

In the nucleic acid delivery literature, the formation of the lipid-nucleic acid complex is compared to or directly linked to the formation of polyelectrolyte-surfactant complexes (PESCs). Nucleic acids are polymers of negatively charged nucleotide monomers linked to a sugar-phosphate backbone analogous to a strongly anionic polyelectrolyte [6]. Similarly, cationic ionizable lipids are amphiphilic and can become positively charged in an aqueous environment, analogous to cationic surfactants [1,6]. Research on ion pairing between polyelectrolytes (PEs) and surfactants has continued since the late 1940s [7–9]. Once combined in solution, like lipid-nucleic acid systems, oppositely charged PEs and surfactants will associate and, under specific conditions, form colloidally or thermodynamically stable nanosized polyelectrolyte-surfactant complexes (PESCs) between 40 and 400 nm [10–13]. The fundamental mechanisms of and parameters that influence self-assembly in polymer-surfactant systems have been well studied, giving way to novel innovations in soft material science with applications in drug delivery [14,15], solar cell technology [16], and water purification [17,18]. As with RNA-LNP complexes, the size and morphology of PESCs are governed by the balance of hydrophobic and electrostatic interactions at the time of complexation [10]. Literature reviews on PESCs indicate that several factors determine the magnitude of these interactions: the physiochemical properties and stoichiometric ratios of the individual components, the composition of the bulk media, and the mixing procedure used to combine polyelectrolyte and surfactant stock solutions [10,19,20]. Compared to the RNA and cationic ionizable lipids used in RNA LNPs, anionic polyelectrolytes and cationic surfactants are affordable, water-soluble, and easily accessible in large quantities. This, paired with the similarity between physiochemical properties and complexation mechanisms, suggests that polyelectrolyte-surfactant complexes are a promising model for studying factors that influence the complexation event during LNP formation.

A critical factor that bridges both fields is the mixing procedure used to combine the aqueous solutions of charged components. Mixing efficiency determines the local concentration of each component and, therefore, the local ratio between components during complexation. Well-studied mixing approaches for PESC systems include drop-wise mixing [21], stop-flow mixing [22], and microfluidic mixing [23]. It is evident from these studies that when the mixing procedure fails to rapidly homogenize PE and surfactant solutions, as in drop-wise mixing, localized differences in component concentrations are present during the complexation [21]. These localized differences modulate the balance between the hydrophobic and electrostatic interactions that govern the size and morphology of the PESC and result in a heterogenous, more polydisperse PESC population [21]. As a result, direct correlations between initial formulation parameters (charge ratio, solution ionic strength, absolute concentration of PE, and surfactant) and the properties of the resulting PESC are challenging to determine systematically without a reproducible mixing platform that rapidly homogenizes the system on a time scale shorter than self-assembly.

Regarding LNPs, many lab-scale studies continue to rely on manual mixing and do not explicitly consider mixing effects. The body of review literature around RNA LNPs has begun to note how the method of LNP preparation can influence the self-assembly process and internal structure of the resulting LNPs [24–26]. For example, the relative ratio between fluid volumes during mixing influences lipid supersaturation ratios and therefore the final size of the LNPs [4]. Most of the existing literature that studies the effects of mixing factors on RNA LNP formulation has focused on microfluidic mixing, which offers a highly reproducible and materially non-intensive platform for mixing [27]. For example, a recent study compared the *in vivo* delivery efficiency of LNPs prepared via microfluidic mixing versus manual mixing and found that optimizing microfluidic mixing parameters was necessary to maximize delivery performance [28]. However, microfluidic mixing processes have inherently poor scalability and are not representative of processes used in the global manufacture of RNA LNPs. Turbulent impinging flow mixing for nanoparticle formulation, developed and termed Flash NanoPrecipitation (FNP) by Prud'homme and collaborators, is an attractive approach for producing LNPs at large scales. FNP is a rapid mixing platform that homogenizes inlet streams on a time scale of ~ 1–5 ms [29], which enables rapid and reproducible mixing on a time scale shorter than particle assembly time [30]. FNP mixers include two main geometries, a confined impinging jet mixer with two inlets and a multi-inlet vortex mixer (MIVM) with four inlets [31]. Pfizer-BioNTech utilizes turbulent flow impinging jet mixers to produce the Comirnaty vaccine [3].

In this study, we utilized the MIVM FNP mixer to isolate and mechanistically study the effects of chemical factors (PE/surfactant stoichiometric ratio, PE and surfactant absolute concentration, etc.) and mixing factors (flow rate, flow fluctuation) on polyelectrolyte-surfactant complexation and LNP formulation [32–34]. Polyelectrolyte-surfactant complexation was studied using the anionic polyelectrolyte polystyrene sulfonate (PSS) and the cationic surfactant cetrimonium bromide (CTAB), which were selected because both their individual properties and fundamental mechanisms of assembly have been well investigated [35–37]. After encapsulation, LNPs are quenched and subjected to an annealing step to preserve the desired LNP structure [25]. Therefore, we studied the characteristics of these complexes immediately after precipitation to capture the system behavior on a time scale representative of RNA encapsulation during LNP formation. In agreement with previous work, we found PESC size to be a function of stoichiometric ratio, absolute concentration, and total volumetric flow rate [19,38,39]. We then demonstrate the sensitivity of this system to mixing factors by producing PESCs under fluctuating flow. This systematic investigation of the sensitivity of polyelectrolyte-surfactant complexation to flow fluctuation is explained as a function of the isolated parameters from the initial experiments. We then replicate these findings using a typical RNA LNP formulation under the same flow fluctuation conditions. This study validates the importance of developing mixing process controls to ensure reproducible and controlled complex formation in PESC and LNP systems at industry scale.

2. Materials and methods

2.1. Materials

Poly(sodium 4-styrenesulfonate) (PSS, MW~1,000,000, ≤10 % moisture, Lot# MKCS3087) was purchased from Sigma-Aldrich (St. Louis, MO, USA). Hexadecyltrimethylammonium bromide (CTAB, ≥98 % purity, Lot# SLCM0580) was purchased from Sigma-Aldrich (St. Louis, MO, USA). Ethanol (USP grade) was purchased from Deacon Labs, Inc. (Prussia, PA, USA). Purified Torulla Ambion™ yeast RNA was purchased as a 10 mg/mL solution from Thermo Fisher Scientific (Waltham, MA, USA). 6-((2-hexyldecanoyl)-N-(6-((2-hexyldecanoyl) oxy)hexyl)-N-(4-hydroxybutyl)hexan-1-aminium (ALC-0315, >99 % purity) was purchased from Avanti Polar Lipids (Alabaster, Alabama, USA). 1,2-dimyristoyl-*rac-glycero*-3-methoxypolyethylene glycol-2000 (DMG-PEG 2000, >99 % purity) was purchased from Avanti Polar Lipids (Alabaster, Alabama, USA). 1,2-disteroyl-*sn-glycero*-3-phosphocholine (18:0 PC (DSPC), >99 % purity) was purchased from Avanti Polar Lipids (Alabaster, Alabama, USA). Cholesterol (>99 % purity) was purchased from Sigma-Aldrich (St. Louis, MO, USA). OmniPur 4-(2-hydroxyethyl)-1-piperazineethanesulfonic acid (HEPES Free Acid, ≥99 % purity) was purchased from Sigma-Aldrich (St. Louis, MO, USA). Quant-it™ RiboGreen RNA Assay Kit was purchased from Thermo Fischer Scientific (Waltham, MA, USA). Sodium hydroxide pellets (NaOH, Lot# 219185, >97 % purity) were purchased from Thermo Fischer Scientific (Waltham, MA, USA). Sodium chloride (NaCl, Lot# 217023, >99 % purity) was purchased from Thermo Fischer Scientific (Waltham, MA, USA). Triton® 100-X was purchased from Thermo Fischer Scientific (Waltham, MA, USA). Water was MilliQ (MQ) grade.

2.2. Preparation of polyelectrolyte-surfactant complexes

Aqueous stock solutions of PSS and CTAB were prepared individually by adding MQ water to measured amounts of either PSS or CTAB, and heating to ~ 50 °C to fully dissolve the CTAB. Solutions were subsequently allowed to cool to room temperature. A complete list of stock solution concentrations and respective experiments is provided in the supplemental information. PSS-CTAB complexes were formed by Flash NanoPrecipitation (FNP) using a multi-inlet vortex mixer (MIVM, Holland Applied Technologies) and two Harvard Apparatus PHD ULTA Satellite – I/W syringe pumps controlled by the Harvard Apparatus pump controller system (HPAC) [40]. The total time per run in each experiment was 16 s, with a sample collection time of 10 s. The mixer effluent produced for the first three seconds and final three seconds of each run was sent to waste. Each PHD ULTA Satellite – I/W syringe pump held two 50 mL glass SGE Analytical Science syringes containing either the PSS or CTAB stock solution for that experiment. Once a run was initiated, the PSS and CTAB stock solutions were rapidly injected into the MIVM, as described elsewhere [40,41]. Upon mixing, PSS and CTAB formed colloidally stable complexes due to a combination of hydrophobic and electrostatic effects [13]. To simulate fluctuating flow conditions, the HPAC was programmed to pulse between different pairs of flow rates that oscillated around a flow rate of 40 mL/min per stream (used for the 'base case' described below). To further examine the effects of fluctuations in flow on the complexation of PSS and CTAB, the flow rate experiments were repeated using peristaltic pumps (Masterflex, Cole-Parmer Instrument Co.). An ultrasonic illumination flow sensor (1PXN, Transonic) with a relative accuracy of ± 0.8 mL/min was used to measure and validate the output flow rates by both the syringe and peristaltic pumps. Flow data were recorded using a flow module (TS410 Flow Module, Transonic) connected to a data acquisition board (X Series Multifunction DAQ, National Instruments). A 1 mL aliquot from the collection vessel was added to a cuvette for size distribution analysis using dynamic light scattering at a 173° backscatter (Malvern Zetasizer Pro, Malvern Instruments). Standard cumulants analysis was used and sample correlograms are provided in the SI. All measurements were performed in triplicate immediately after precipitation to capture the initial dynamic state of these non-equilibrium systems as discussed in the introduction. Error bars in all figures are representative of the standard deviation of three separate runs using the same stock solutions.

2.3. Base case

A set of specific chemical and mixing conditions, referred to as the 'base case,' were used as a baseline for comparison among the different isolated chemical and mixing factors studied herein. In this base case, a stoichiometric charge ratio (1:1 CTAB:PSS) was used; this refers to the ratio between the moles of cationic nitrogen in the CTAB headgroup and the moles of anionic sulfonate groups on the backbone of PSS as indicated by the red circles in the top of Fig. 1.

The concentration of PSS in the base case formulation was fixed to 1 mg/mL, and a stock solution of CTAB was prepared at 2.13 mg/mL to achieve the 1:1 CTAB:PSS charge ratio. Therefore, the total concentration of dissolved solids in the mixing chamber, referred to in this work as total solids concentration (TSC), was 1.57 mg/mL in the base case. The flow rate of each inlet stream was set to 40 mL/min, generating a total flow rate of 160 mL/min, which has been demonstrated to be sufficient to achieve homogenous mixing on a time scale of 1.5–3 ms in the MIVM [40,41]. Total volumetric flow rate and relative inlet stream flow rate are not parameters traditionally explored in PESC literature but have been reported as significant in previous literature utilizing FNP mixing [42].

2.4. Preparation of lipid nanoparticles

Lipid nanoparticle preparation involved three components: (1) a collection bath, (2) an aqueous feed stream solution containing RNA, and (3) an ethanol feed stream containing the lipid mixture. First, a 100 mM solution of (4-(2-hydroxyethyl)-1-piperazineethanesulfonic acid) (HEPES) buffer was prepared and then adjusted to the desired pH of 7 (final pH=7.00) using 1 M NaOH. This 100 mM solution was diluted using MQ water to yield a 10 mM HEPES buffer solution and a 20 mM HEPES buffer solution. The 10 mM solution was utilized as the quench bath to collect the effluent of the mixer after nanoprecipitation, reducing the ethanol volume fraction to 10 % from 50 % and arresting the growth of the LNPs (1). Yeast RNA was added to the 20 mM HEPES buffer solution to yield a total RNA concentration of 0.6 mg/mL (2). The lipid mixture in ethanol (3) was prepared at a molar ratio of 50:10:38.5:1.5 (ALC-0315:DSPC:cholesterol:DMG-PEG), the same ratio utilized in the Onpattro (Patisiran) siRNA LNP formulation developed by Alnylam and the mRNA-1273 (Spikevax) mRNA LNP developed by Moderna [26,43]. This correlated to an N/P ratio (charge ratio between the nitrogen on the cationic ionzable lipid ALC-0315 and the phosphate on the RNA nucleotides) of 6:1. The final lipid feed solution contained a total dissolved lipid concentration of 12 mg/mL.

The same stock solutions were fed into the MIVM using syringe pumps or peristaltic pumps to produce lipid nanoparticles. Operation of the syringe pumps was as previously mentioned. One syringe pump was loaded with two 50 mL SGE syringes containing the 0.6 mg/mL RNA solution in 20 mM HEPEs buffer. The second syringe pump had two 25 mL SGE syringes containing the lipid stock solution. The syringe pumps were programmed to push each syringe at 40 mL/min for a total volumetric flow rate equal to 160 mL/min. The total time run time in this experiment was 16 s with sample collection time equal to 10 s (three seconds sent to waste at the start and end of each run). The effluent of

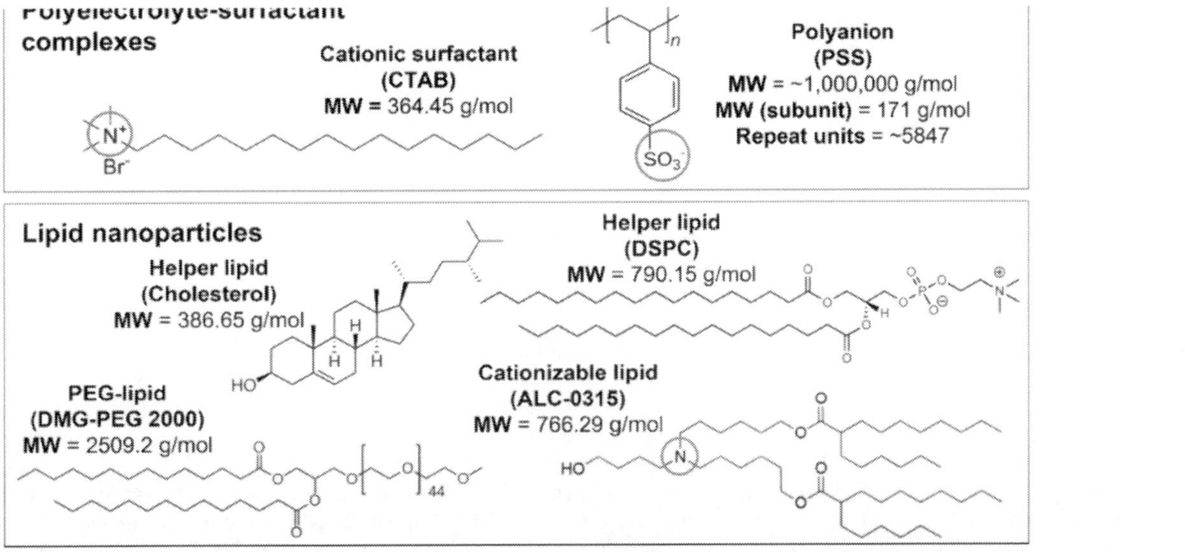

Fig. 1. (Top) Chemical structures and relevant molecular weights of the polyanion (Polystyrene sulfonate, PSS) and cationic surfactant (cetrimonium bromide, CTAB) used to produce polyelectrolyte-surfactant complexes in this work. (Bottom) Chemical structures and molecular weights of the helper lipids (Cholesterol, DSPC), cationic ionizable lipid (ALC-0315), and PEG-lipid (DMG-PEG) used to produce lipid nanoparticles. Ionizable groups that undergo electrostatic complexation are circled in red. (For interpretation of the references to colour in this figure legend, the reader is referred to the web version of this article.)

the mixer was collected into a 270 mL quench bath containing pH 7 10 mM HEPEs buffer solution to dilute the ethanol volume fraction to 10 %.

Prior to the experimental run using the peristaltic pumps, the flow rate of each individual peristaltic pump was validated. Each peristaltic pump was set to a flow rate of 80 mL/min such that, after being split by a *t*-split downstream, the inlet flowrate was 40 mL/min per stream, matching the syringe pumps. One peristaltic pump inlet contained the 0.6 mg/mL RNA solution in 20 mM HEPEs buffer, and the second peristaltic pump inlet contained the lipid feed solution. The total time run time in this experiment was 25 s with sample collection time equal to 10 s. The effluent of the mixer was collected into a 270 mL quench bath containing pH 7 10 mM HEPEs buffer solution. The size and surface charge of the resulting lipid nanoparticles produced using both systems were measured using the Malvern Zetasizer Pro (Malvern Instruments).

2.5. *Encapsulation efficiency of the lipid nanoparticles*

The Quant-it RiboGreen RNA Assay was used to determine the encapsulation efficiency of RNA in the lipid nanoparticles produced using both types of pumps. Prior to running the assay, 1x Tris EDTA buffer was prepared by diluting the 20x Tris EDTA buffer provided in the kit. Additionally, 0.2 g of Triton® 100-X was added to 10 mL of 1x Tris buffer and allowed to stir overnight. The resulting 2 % Triton® X-100 solution was used in this assay to disrupt the lipid nanoparticle structure and to facilitate the quantification of the total RNA in the sample. To run the assay, a 96-well plate was loaded with three replicates of an LNP sample from each experiment prepared according to the assay instructions both with and without Triton® 100-X (12 total wells) and two calibration curve samples prepared using the kit RNA with and without the Triton® 100-X (done in duplicate, 28 total wells). All samples on the plate contained the RiboGreen dye. Assay samples were prepared in microcentrifuge tubes and then added to the plate in 100 uL aliquots. Empty wells were filled with 100 uL MQ water. Once the RiboGreen dye was added to the sample, the samples were homogenized via gentle mixing and incubated for 10 min prior to measurement. Upon binding to RNA, the RiboGreen reagent excites at a wavelength of 485 nm and fluoresces green at a wavelength of 528 nm. The SpectraMax i3x by Molecular Devices (San Jose, CA) was used to record the intensity of the emitted light. The relative intensity of the emitted light was correlated to the concentration of RNA in the sample using the appropriate calibration curve. Encapsulation efficiency was then calculated using the relative concentration of RNA in a sample with and without Triton® 100-X: $EE\% = 100*(1 - \frac{C_{without\ Triton®\ 100-X}}{C_{with\ Triton®\ 100-X}})$.

3. Results and discussion

3.1. *MIVM orientation effects*

There are four possible inlet stream arrangements on an MIVM: 'combined adjacent,' 'combined opposite,' 'separate adjacent,' and 'separate opposite' [40]. In this work, it is desired for the stream components to remain separate until they are mixed in the chamber, limiting the acceptable orientations to separate opposite or separate adjacent. Given the sensitivity of PESCs to mixing protocol, it was important to determine if there was a PESC size dependence on orientation. The PSS-CTAB complexes were 75.0 ± 0.9 nm using the separate opposite arrangement and 73.9 ± 0.9 nm using the separate adjacent arrangement. These sizes show good agreement with one another, which is further supported by the particle size distributions provided in the SI. This agreement suggests that complex size is independent of MIVM inlet stream arrangement in this case. The 'separate opposite' arrangement was utilized herein for consistency across experiments [16,17].

3.2. *Chemical factor: Total solids concentration*

Polyelectrolyte-surfactant complexation literature investigating the effects of component concentration tends to fix either the concentration of surfactant or the concentration of polyelectrolyte, and modulate the concentration of the other component [44]. As a result, these studies simultaneously vary both the charge ratio and total solids concentration during complexation rather than isolating each parameter. To decouple the effects of charge ratio and TSC, we prepared PESCs with total solids concentrations between 0.20 mg/mL and 3.13 mg/mL at a fixed charge ratio of 1:1 CTAB:PSS. We fixed the flow rate parameters to be identical to those in the base case. We explain our findings, presented in Fig. 2A, using the established relationship between total solids concentration and colloid size for self-assembled polymer-stabilized nanoparticles

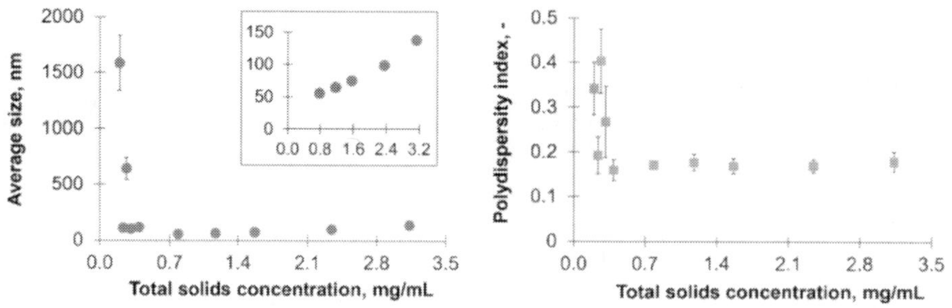

Fig. 2. (A) Complex size and (B) polydispersity index (PDI) as a function of the dissolved total solids concentration (TSC). Complexes in this series of experiments were produced at a 1:1 CTAB:PSS charge ratio and at a total volumetric flow rate of 160 mL/min in line with the base case conditions. These results demonstrate that PSS-CTAB complexes produced when CTAB is below the CMC are large and polydisperse compared to complexes produced when CTAB is above the CMC. Error bars on both plots and 2(A) inset represent the standard deviation of n = 3 runs. There are some data points where the marker size is larger than the plotted error bars.

produced using Flash NanoPrecipitation [42,45].

Above a total solids concentration of 0.4 mg/mL, polyelectrolyte-surfactant complex size increased with increasing total solids concentration, in line with expectations from FNP literature [42]. In the 0.20 to 0.39 mg/mL range, complexes demonstrated increasing size and polydispersity with decreasing total solids concentration, contradictory to expected results. This may be attributable to the concentration of CTAB being below its critical micelle concentration (CMC). It has been reported that the formation of homogenous, densely packed PSS-CTAB structures is largely due to the solubilization of the aromatic sulfonate rings on the PSS backbone by CTAB micelles, maximizing hydrophobic interactions [36,39]. However, when the concentration of CTAB is below its CMC (which is reported to be ~ 0.32 mg/mL in deionized water), only free surfactant molecules are present in solution. In such a case, no micelles are present to sufficiently solubilize the PSS backbone [46]. In this experiment, the concentration of CTAB is below or near the reported CMC at every point in the range where we observed the trend of increasing PESC size with decreasing TSC. Furthermore, the presence of PSS in solution has been demonstrated to increase the CMC of CTAB, supporting the hypothesis that there is insufficient micellization of CTAB in this range [34]. Thus, PSS-CTAB complexes do not show agreement with expected trends until the concentration of CTAB in the mixing chamber is well above the CMC at ~ 0.54 mg/mL (TSC=0.78 mg/mL).

3.3. *Chemical factor: Charge ratio*

To isolate the effect of charge ratio on PESC formation, complexes were prepared at CTAB:PSS charge ratios of 1:4, 1:2, 1:1, 2:1, and 4:1 at three different fixed total solids concentrations (0.78 mg/mL, 1.57 mg/mL, and 3.13 mg/mL). The volumetric flow rate was fixed across experiments to 160 mL/min in agreement with the base case. Consistent with findings by Pojják et al. [47] (who utilized non-scalable but rapid stop-flow mixing to interrogate PSS-CTAB complexation in a comparable range of charge ratios, but with different total solids concentrations at each point) and others [48,49], PESC size reaches a maximum at the stoichiometric charge ratio (1:1 CTAB:PSS). When either the PSS (e.g. CTAB:PSS charge ratios of 1:2, 1:4) or CTAB (e.g. CTAB:PSS charge ratios of 2:1, 4:1) is in stoichiometric excess, complexes decrease in size relative to the 1:1 charge ratio. We additionally observe that complexes formed at higher total solids concentrations are larger, consistent with the results demonstrated above.

We present these results in Fig. 3.

The trend observed in Fig. 3B regarding the polydispersity of PSS-CTAB complexes when PSS is in excess may be attributed to the complexes' morphology. A study investigating complexes formed by PSS and tetradecyltrimethylammonium bromide (TTAB) as a function of charge ratio [50] found that as the ratio of TTAB to PSS increased, the PSS chain swelled and coiled in on itself, eventually undergoing compaction [50]. A similar increase in coil contraction was established for PSS-CTAB complexes using complex size measurements as a function of increasing CTAB concentration [36]. We hypothesize that this difference in morphology explains the polydispersity seen when PESCs are formed under reproducible mixing. When surfactant is in excess, PESCs formed are more densely packed with lower PDIs than complexes formed in excess polyelectrolyte (Fig. 3B) [36]. Since DLS measurements assume rigid spheres, more densely-packed PESC samples are expected to generate more uniform DLS readings (with lower PDI) than those with less dense packing, i.e. when polyelectrolyte is in excess [51]. It is

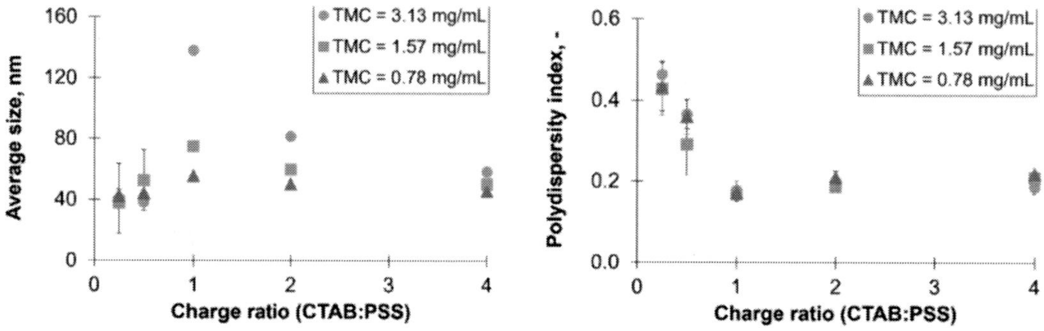

Fig. 3. (A) Complex size and (B) polydispersity index (PDI) as a function of charge ratio at three fixed total dissolved solids concentrations (TSC). The base case TSC, 1.57 mg/mL, is represented by the orange squares. All complexes were produced at a total volumetric flow rate of 160 mL/min, in line with the base case. For all TSCs, it is observed that complex size reaches a maximum at a charge ratio of 1:1 (CTAB:PSS). Error bars represent the standard deviation of n = 3 separate runs. There are some data points where the marker size is larger than the plotted error bars. (For interpretation of the references to colour in this figure legend, the reader is referred to the web version of this article.)

important to note that at all data points collected in this series, the concentration of CTAB was above its CMC, meaning the variation in polydispersity observed in this series is more likely attributable to the final PESC morphology rather than insufficient hydrophobic interactions (the explanation given for the increase in PDI seen in Fig. 2).

The results presented so far demonstrate that the rapid mixing provided by FNP can reliably produce PSS-CTAB complexes that recapitulate trends reported in the literature for the chemical factors of total solids concentration and charge ratio. These results validate the platform as a method for isolating and studying mixing factors such as volumetric flow rate, relative stream flow rates, and fluctuating flow, which follow.

3.4. Mixing factor: Volumetric flow rate

Volumetric flow rate correlates with Reynolds number, a key parameter studied in FNP literature to ensure that a homogeneous system is achieved on the order of a few milliseconds, eliminating the effects of localized concentration differences. It has been established that a Reynolds number in the range of ~ 20,000–25,000 (total volumetric flow rate of 160 mL/min) is required to achieve this degree of mixing on the appropriate time scale in an MIVM [40,41]. Below this threshold, slower mixing is expected to result in inhomogeneous PESC assembly due to local concentration gradients [22]. To demonstrate this relationship, PSS-CTAB complexes were produced using total volumetric flow rates between 40 mL/min (Re = 7,500) and 320 mL/min (Re = 60,200). This range is in agreement with the Reynolds number regime specified for the design of the MIVM used (see Fig. 4a in Markwalter et al.) [40]. Results of this series are provided in Fig. 4. Consistent with other FNP literature, complex size decreases with increasing Re to a threshold around ~ 25,000, above which no significant dependance of PSS-CTAB complexation on Re is observed.

In addition to total volumetric flow rate, we studied the effects of varying the individual inlet stream flow rates of the PSS and CTAB streams. In this experimental series, the inlet stream concentrations were determined such that the base case conditions would be replicated at the time of complexation in the mixing chamber despite the mismatched inlet stream flow rates. The sum of the relative inlet stream flow rates across all four streams was equal to a total volumetric flow rate of 160 mL/min, in line with the base case. The results are presented in Fig. 5. Little difference in particle size and polydispersity (measured values listed in the SI) is observed. This indicates that the size and formation of PSS-CTAB complexes are independent of the relative stream flow rates input into the MIVM because the desired concentrations are maintained at the time of complexation and the total flow rate is sufficient for good mixing. This is particularly significant for systems using mixed aqueous/organic solvents, which rely on supersaturation of components in the mixed solvent in the mixing chamber.

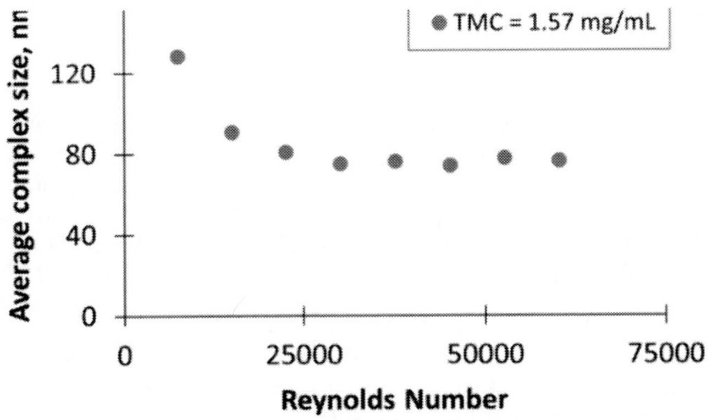

Fig. 4. PSS-CTAB complex size dependence on Reynolds number (Re) when complexes are produced via Flash NanoPrecipitation. The base case formulation parameters were used in each run. Above a Re of ~ 25,000, there is no significant correlation between complex size and Reynolds number. Below this threshold Re, as Re decreased, complexes increased in size and were more polydisperse. Error bars represent the standard deviation of three runs. All error bars in this series are smaller than the markers and represent the standard deviation of n = 3 individual runs.

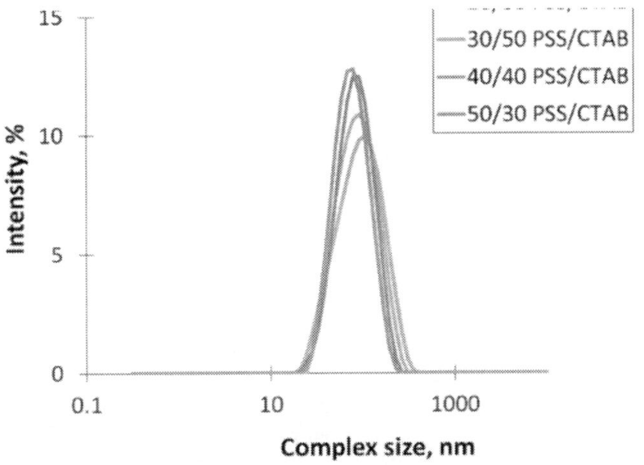

Fig. 5. Particle size distributions of PSS-CTAB complexes produced via Flash NanoPrecipitation at different combinations of inlet volumetric flow rates. In all cases, the total dissolved solids concentration (TSC), charge ratio, and total volumetric flow rate were fixed to be the base case the mixing chamber at the time of complexation (1.57 mg/mL TSC, 1:1 CTAB:PSS, and 160 mL/min). The numbers in the figure legend correspond to the individual volumetric flow rates of PSS and CTAB feed streams in mL/min. For example, the orange curve represents the case where the PSS inlet streams were set to a flowrate of 20 mL/min each and the CTAB inlet flow streams were set to 60 mL/min each, for a total inlet volumetric flow rate of 160 mL/min. Each distribution represents the average of n = 3 individual runs. (For interpretation of the references to colour in this figure legend, the reader is referred to the web version of this article.)

3.5. Mixing factor: Simulated fluctuating flow

Here we aimed to demonstrate the effect of flow fluctuation on PESC assembly. Steady flow is recommended in the FNP literature [31], but the exact effect of flow fluctuations on nanoparticle or PESC assembly under FNP mixing have not been reported before. Our hypothesis was that, since fluctuating flow may affect the above parameters at a given time – total flow rate, relative stream flow rate, total solids concentration in the mixing chamber, and charge ratio – unsteady flow would result in polydisperse particle populations. We studied the effects of flow fluctuation by investigating three unique simulated cases of pulsatile flow and then by producing PSS-CTAB complexes using a pair of fluctuating peristaltic pumps. In all studies, the base case stock solution concentrations of PSS and CTAB were utilized. In each of the simulated flow cases, the magnitude of flow fluctuation (deviation from the base case of 40 mL/min per stream) was varied from 5 % to 50 % (±2 mL/min to ± 20 mL min) at a constant frequency of oscillation equal to two seconds. This periodicity was determined from previous experiments and was found to be sufficient for (1) the syringe pumps to reach the extrema for the 50 % magnitude case and (2) capture the effects of unsteady mixing on PESC formation. Each of these experiments was performed in triplicate with the measured size and polydispersity index values plotted in Fig. 6(D–F). Flow sensor data for each of these simulated cases is provided in Fig. 6(A–C). An analog of Fig. 6 using Reynolds number in the y-axis of panels (A-C) is provided in the SI as Fig. S2.

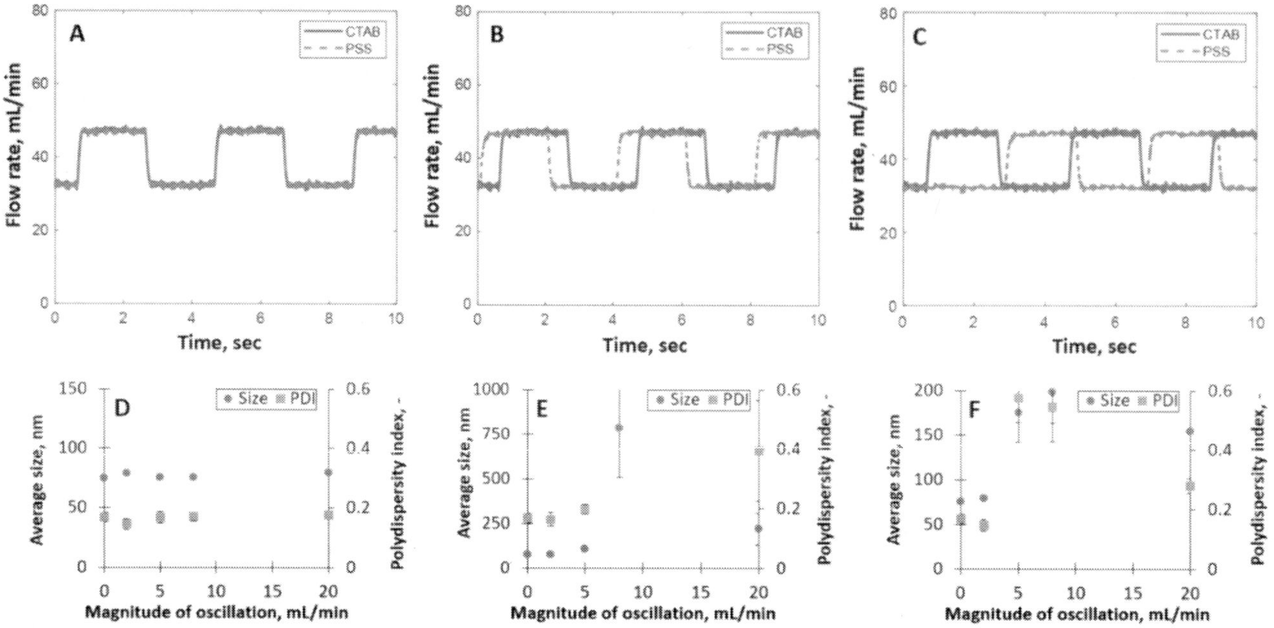

Fig. 6. (A–C) Flow meter output for each of the three types of flow fluctuation cases programmed into the syringe pumps for a fixed oscillation amplitude of ± 8 mL/min. The orange dotted line represents the flow rate of the PSS streams, and the blue solid line represents the flow rate of the CTAB streams. In Case I (A), the inlet streams oscillate around the base case flow rate in phase with one another. In Case II (B), the inlet streams oscillate around the base case flow rate out of phase with one another. Case III (C) investigates an intermediate condition between Cases I and II. (For interpretation of the references to colour in this figure legend, the reader is referred to the web version of this article.)

3.5.1. *Case I: In-phase fluctuation*

When the syringe pumps were programmed to vary in phase with one another (Fig. 6A), no major change in particle size or PDI was observed (Fig. 6D). The total solids concentration and charge ratio at each condition were calculated and found to be constant at 1.57 mg/mL and 1:1 (CTAB:PSS), respectively. The only change across a given run when the syringe pumps were fluctuating in phase were the minimum and maximum total volumetric flow rates during a run, which reached a maximum of 240 mL/min (Re ~ 45,180) and minimum of 80 mL/min (Re ~ 15,060) at the most extreme case of ± 20 mL/min fluctuation. The PSS-CTAB complexes produced under this extreme case were 79.64 ± 1.1 nm. From Fig. 4, we expect that the PSS-CTAB complexes produced at a Re ~ 45,180 and Re ~ 15,060 to represent sizes between 73.9 ± 0.5 nm and 90.7 ± 1.4 nm. The result of 79.64 ± 1.1 nm shows good agreement with this expectation.

3.5.2. *Case II: Out-of-phase fluctuation*

In this case, a range of total solids concentrations and charge ratios were generated in the mixing chamber during a given run. This significantly compromised PESC complexation, captured by the variety of sizes and polydispersity of the complexes presented in Fig. 6E. The results presented in Fig. 5 demonstrated that different relative volumetric flow rates alone are not enough to generate heterogeneous populations of PESCs that differ significantly from the base case. Therefore, we can attribute these findings to the variety of total solids concentrations and charge ratios generated in the mixing chamber within the same run.

For example, in the run programmed to have the most extreme magnitude of oscillation (±20 mL/min from the base case), the total solids concentrations at the time of complexation varied from 0.78 mg/mL to 2.35 mg/mL and the charge ratio varied from 0.3 to 3 (CTAB:PSS). From the results presented in Fig. 3A, it is expected that PSS-CTAB complexes produced at 0.78 mg/mL and a charge ratio of 3:1 (CTAB:PSS) are ~ 50–40 nm while complexes produced at 2.35 mg/mL and a charge ratio of 1:3 (CTAB:PSS) are of comparable size but significantly more polydisperse. Compared to these predictions, the average complex size reported by the DLS (219.8 ± 89.1 nm) for this experiment was large. This discrepancy can be attributed to the fact that the extrema sample points isolated for comparison across experiments are only a small portion of the full set of conditions generated during this experiment. Each of the samples reported in Fig. 6E is, therefore, a heterogeneous population of complexes generated under a variety of conditions. For all experiments reported in this section, the Malvern software on the DLS instrument indicated that there were multiple populations detected in the sample, with higher polydispersity indices and larger sizes than the base case. The apparent size/PDI maximum at ± 7 mL/min fluctuation (especially compared to the ± 20 mL/min fluctuation) may be attributable to the cumulants fit failing to resolve highly polydisperse populations. These results emphasize the sensitivity of this system (and FNP in general) to fluctuating flow as an increase in size and PDI are observed even for deviations from the base case as low as ± 2 mL/min, a 5 % fluctuation.

3.5.3. *Case III: Partially in-and-out-of-phase fluctuation*

In Case III a large range of unique total solids concentrations and charge ratios were also generated in the mixing chamber. However, due to the initial offset of the inlet stream flow rates, the minimum and maximum values for these conditions were not as extreme as Case II. As a result, the concentration gradients of PSS and CTAB in the mixing chamber at any given time varied less and thus the PSS-CTAB complexes produced within a given run, although heterogenous, were not as polydisperse as the complexes produced in Case II. We present these results in Fig. 6F. The decreased magnitude of the error bars in Fig. 6F compared to Fig. 6E suggests that the complexes produced when flow fluctuation is only partially out of phase are more reproducible across runs than those produced when flow fluctuation is entirely out of phase.

3.6. *Mixing factor: Pulsatile flow with peristaltic pumps*

Peristaltic pumps generate oscillatory flow patterns as an artifact of their operating mechanism [52]. The oscillatory nature of peristaltic pumps was exploited in this work as a comparison to the oscillatory flow simulated by the syringe pumps to further elucidate the effects of flow fluctuation on polyelectrolyte surfactant complexation. First, we demonstrated the oscillatory behavior of peristaltic pumps by taking flow rate measurements over time using a flow sensor and MQ water. The results are presented in Fig. 7A and compared to the steady flow behavior of the syringe pumps in Fig. 7B. From these measurements, we can conclude that the oscillatory behavior of a peristaltic pump is dependent on flow rate with the frequency and magnitude of the oscillations increasing with increasing flow rate.

3.7. Lipid nanoparticles produced under pulsatile and steady flow conditions

To further demonstrate the compromising effects of localized concentration fluctuation on the formation of ion paired systems, lipid nanoparticles (LNPs) were produced using syringe pumps and peristaltic pumps. From the previously presented results, we hypothesized that the LNPs produced under pulsatile flow conditions would have wider particle size distributions and lower encapsulation efficiencies. We present these results in Fig. 8(C&F) alongside the results for PESCs produced under identical conditions (B&E).

The LNPs produced using peristaltic pumps had monomodal particle size distributions with narrow PDIs comparable to the LNPs produced using syringe pumps. The most notable differences between the two preparation methods are the particle size and encapsulation efficiency (EE). Specifically, the LNPs produced with the peristaltic pumps were around twice the size of the LNPs produced using steady flow conditions and had a significantly lower encapsulation efficiency. This difference in EE may be attributed to the importance of maintaining the charge ratio between the RNA and cationic ionizable lipids. This pre-determined stoichiometric ratio is necessary to precipitate the RNA for encapsulation and is used in practice to modulate the quantity of RNA to be encapsulated [25]. The difference in size between the two LNP samples may also be attributed to the influence of flow fluctuations on LNP self-assembly. LNPs have a core–shell architecture that is is largely a function of the initial molar ratio between lipids, and the liquid crystalline mesophase core structures are a function of the chemical structures of those lipid components [5,25]. Even under fluctuating flow, the molar ratio of lipids is held constant since all lipids are co-dissolved in the ethanol feed stream. What does vary under fluctuating flow, though, is the relative volume ratio between ethanol and water in the mixing chamber, which controls the degree of supersaturation for the lipids. These different degrees of supersaturation will produce LNPs larger and smaller than in the case for the steady flow conditions where the inlet flow rate and, thus, ethanol–water volume ratio in the chamber are fixed. Some overlap of the two distributions is expected as the conditions in the mixing chamber produced by the peristaltic pumps periodically mirror those in the mixing chamber using the syringe pumps. Ultimately, the structure and loading of LNPs are governed by the molar ratio between the lipids, the stoichiometric ratio between the cationic ionizable lipids and the RNA, and the volumetric ratio between ethanol and water at the time of complexation [4,25]. These ratios are compromised by fluctuating flow, reducing the encapsulation efficiency and affecting the self-assembly process.

4. Conclusions

The results presented here take steps toward a more robust mechanistic understanding of the effect of mixing factors on systems that utilize polyelectrolyte-surfactant complexation and precipitation, including RNA lipid nanoparticles. The rapid and reproducible mixing provided by the FNP platform, which has been utilized at the global scale in the pharmaceutical industry for the formulation of mRNA LNPs, enables individual process variables to be isolated and studied under uniform mixing. FNP is easily scalable unlike other popular mixing platforms used to study LNPs (microfluidics [27]) and ion paired systems (stop flow mixing [22]). The rapid combination of inlet streams unique to FNP eliminates local differences in component concentrations, meaning complexation occurs by diffusion-limited aggregation not achievable by previously studied platforms such as dropwise mixing [21].

The main drawback of utilizing FNP mixing to isolate individual process variables relevant to LNP formulation is its material intensity. The smallest batches that can be prepared using pump-driven flow on commercially-available FNP mixers (including volume sent to waste during startup and shutdown, but not including the subsequent quench used to lower the volume fraction of ethanol below 50 % in this work) are on the order of 15 mL. This is acceptable for systems using low-cost chemicals such as PSS and CTAB but can be prohibitively materially- and cost-intensive with lipids and mRNA. We maintain that the combined approach of using a model system, such as the PSS-CTAB complexes in this study, to identify trends and then replicate selected trends with RNA LNPs is a reasonable compromise.

Our findings using this platform emphasize the importance of steady flow in both polyelectrolyte-surfactant complexation and LNP self-assembly. Oscillatory deviations from steady flow as low as $\pm$ 5 % of total flow are significant enough to compromise mixing and in turn ionic complexation, particle self-assembly, and RNA encapsulation, a critical quality attribute in LNP manufacturing. This effect is captured in both the model system and LNPs. Notably, the LNPs produced using peristaltic pumps had a 49 % decrease in encapsulation efficiency compared

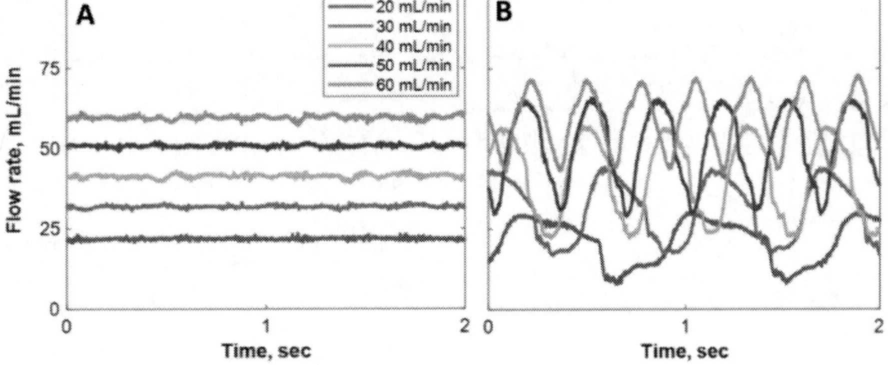

Fig. 7. Flow meter output for (A) syringe and (B) peristaltic pumps for flow rates in the range of 20 mL/min to 60 mL/min. The legend and Y-axis labels in Fig. 7A are applicable to 7A&7B. In this experiment, MQ water was used to obtain the flow rate readings given a comparable density to solutions of dilute inlet stream components. From the peristaltic pump readings, we determine that there is an increase in the frequency and amplitude of the oscillations as flow rate increases.

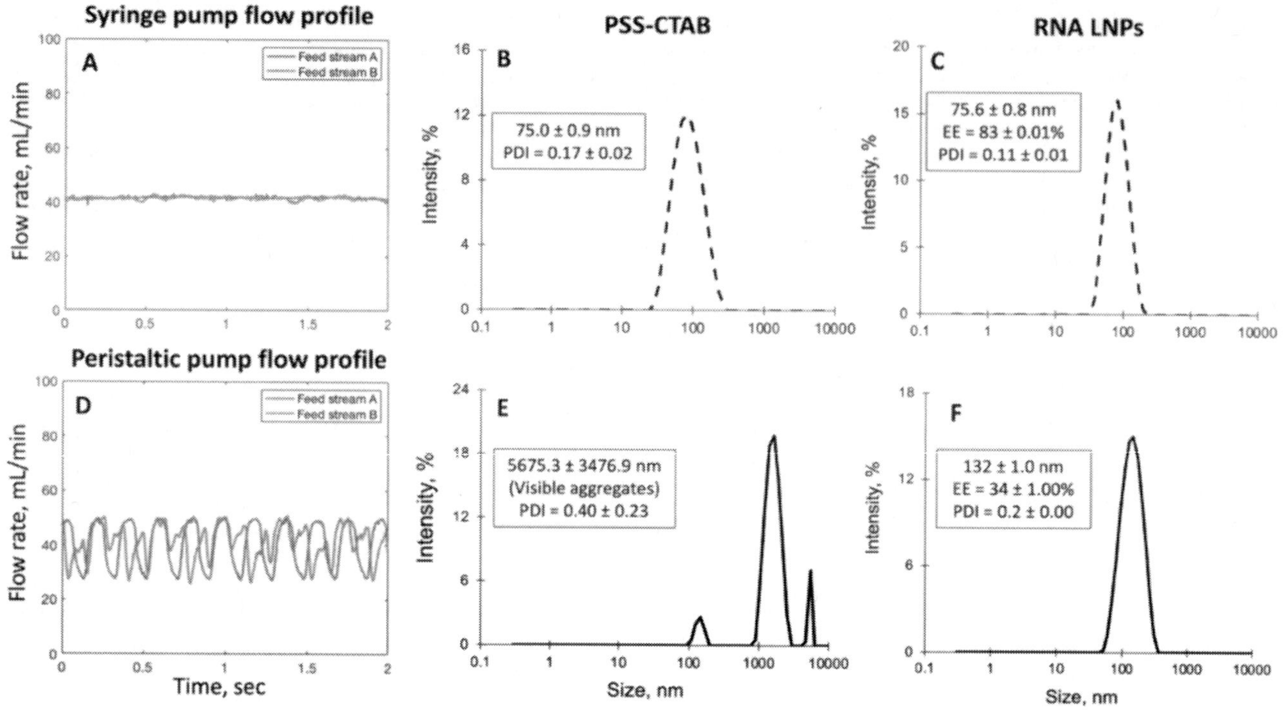

Fig. 8. Flow rate (mL/min) versus time (*sec*) profile at the inlet of the MIVM for flow generated with syringe pumps (A) vs peristaltic pumps (D). (B) and (E) Particle size distribution of PSS-CTAB complexes produced under base case conditions using syringe pumps (B) and the peristaltic pumps (E). (C) and (F) Particle size distribution of the final lipid nanoparticle suspension for nanoparticles produced using the syringe pumps (C) and the peristaltic pumps (F). The average size, polydispersity index (PDI), and encapsulation efficiency (EE) are additionally reported in the boxes where applicable. The reported standard deviations represent the difference between replicate measurements (n = 3) of the same sample for the LNPs and an average of n = 3 runs for the PSS-CTAB complexes.

to the syringe pumps. PESCs produced under the same conditions were polydisperse and lacked long-term colloidal stability.

Future work will utilize this platform to take successive steps toward full LNP relevance by isolating and studying other variables that are expected to affect LNP formulation, such as RNA molecular weight, lipid chemistry, lipid ratio, lipid supersaturation, process temperature, and pH and buffer effects. Other subsequent tests will also utilize *in vitro* and/or *in vivo* protein expression from mRNA as a readout to evaluate formulation quality. This systemic approach to developing a deeper mechanistic understanding of PESC formation and LNP formulation is enabled by the rapid and reproducible mixing afforded by FNP under steady flow, highlighting the importance of directly considering mixing effects in nanoprecipitation.

CRediT authorship contribution statement

Sophia R. Dasaro: Writing – review & editing, Writing – original draft, Visualization, Methodology, Investigation, Data curation. **Abhishek Singh:** Writing – review & editing, Validation, Methodology, Investigation. **Pavlos Vlachos:** Writing – review & editing, Validation, Supervision. **Kurt D. Ristroph:** Writing – review & editing, Validation, Supervision, Resources, Methodology, Funding acquisition, Conceptualization.

Declaration of competing interest

The authors declare that they have no known competing financial interests or personal relationships that could have appeared to influence the work reported in this paper.

Data availability

Data will be made available on request.

Appendix A. Supplementary data

Supplementary data to this article can be found online at https://doi.org/10.1016/j.jcis.2024.08.150.

References

[1] M.G. Kim, S.D. Jo, J.H. Jeong, S.H. Kim, Nanoscale polyelectrolyte complexes encapsulating mRNA and long-chained siRNA for combinatorial cancer gene therapy, J. Ind. Eng. Chem. 64 (2018) 430–437.

[2] R. Gaspar, F. Coelho, B.F. Silva, Lipid-nucleic acid complexes: Physicochemical aspects and prospects for cancer treatment, Molecules 25 (21) (2020) 5006.

[3] A. Sealy, Manufacturing moonshot: How Pfizer makes its millions of Covid-19 vaccine doses, https://www.cnn.com/2021/03/31/health/pfizer-vaccine-manufacturing/index.html (accessed 2021).

[4] Y. Eygeris, M. Gupta, J. Kim, G. Sahay, Chemistry of lipid nanoparticles for RNA delivery, Acc. Chem. Res. 55 (1) (2021) 2–12.

[5] Y. Eygeris, S. Patel, A. Jozic, G. Sahay, Deconvoluting lipid nanoparticle structure for messenger RNA delivery, Nano Lett. 20 (6) (2020) 4543–4549.

[6] Y.H. Kim, K. Lee, S. Li, Nucleic acids based polyelectrolyte complexes: their complexation mechanism, morphology, and stability, Chem. Mater. 33 (20) (2021) 7923–7943.

[7] A. Rembaum, E. Sélégny, Polyelectrolytes and their Applications, Springer Science & Business Media, 2012.

[8] S.M. Shaban, J. Kang, D.-H. Kim, Surfactants: recent advances and their applications, Compos. Commun. 22 (2020) 100537.

[9] A.M. Schwartz, J.W. Perry, Surface active agents, their chemistry and technology, 1949.

[10] B. Lindman, F. Antunes, S. Aidarova, M. Miguel, T. Nylander, Polyelectrolyte-surfactant association—from fundamentals to applications, Colloid J. 76 (2014) 585–594.

[11] B. Lindman, T. Nylander, Polymer-surfactant interactions, in: Cosmetic Science and Technology: Theoretical Principles and Applications, pp. 449–469, 2017.

[12] D. Li, N.J. Wagner, Universal binding behavior for ionic alkyl surfactants with oppositely charged polyelectrolytes, J. Am. Chem. Soc. 135 (46) (2013) 17547–17555.

[13] A.F. Thünemann, Polyelectrolyte–surfactant complexes (synthesis, structure and materials aspects), Prog. Polym. Sci. 27 (8) (2002) 1473–1572.

[14] S. Lankalapalli, V. Kolapalli, Polyelectrolyte complexes: a review of their applicability in drug delivery technology, Indian J. Pharm. Sci. 71 (5) (2009) 481.

[15] M. Gradzielski, Polyelectrolyte–surfactant complexes as a formulation tool for drug delivery, Langmuir 38 (44) (2022) 13330–13343.

[16] Y.M. Chang, R. Zhu, E. Richard, C.C. Chen, G. Li, Y. Yang, Electrostatic self-assembly conjugated polyelectrolyte-surfactant complex as an interlayer for high performance polymer solar cells, Adv. Funct. Mater. 22 (15) (2012) 3284–3289.

[17] Y.V. Shulevich, T.H. Nguyen, D.S. Tutaev, A.V. Navrotskii, I.A. Novakov, Purification of fat-containing wastewater using polyelectrolyte–surfactant complexes, Sep. Purif. Technol. 113 (2013) 18–23.

[18] G. Petzold, S. Schwarz, Dye removal from solutions and sludges by using polyelectrolytes and polyelectrolyte–surfactant complexes, Sep. Purif. Technol. 51 (3) (2006) 318–324.

[19] N. Khan, B. Brettmann, Intermolecular interactions in polyelectrolyte and surfactant complexes in solution, Polymers 11 (1) (2019) 51.

[20] E. Guzmán, A. Maestro, F. Ortega, R.G. Rubio, Association of oppositely charged polyelectrolyte and surfactant in solution: equilibrium and nonequilibrium features, J. Phys.: Condens. Matter, (2023).

[21] A. Naderi, P.M. Claesson, M. Bergstrom, A. Dedinaite, Trapped non-equilibrium states in aqueous solutions of oppositely charged polyelectrolytes and surfactants: effects of mixing protocol and salt concentration, Colloids Surf. A Physicochem. Eng. Asp. 253 (1–3) (2005) 83–93.

[22] A. Mezei, R. Mészáros, I. Varga, T. Gilányi, Effect of mixing on the formation of complexes of hyperbranched cationic polyelectrolytes and anionic surfactants, Langmuir 23 (8) (2007) 4237–4247.

[23] A. Bezrukov, Y. Galyametdinov, Tuning properties of polyelectrolyte-surfactant associates in two-phase microfluidic flows, Polymers 14 (24) (2022) 5480.

[24] M.N. O'Brien Laramy, A.P. Costa, Y.M. Cebrero, J. Joseph, A. Sarode, N. Zang, L. J. Kim, K. Hofmann, S. Wang, A. Goyon, S.G. Koenig, M. Hammel, G.L. Hura, Process Robustness in Lipid Nanoparticle Production: A Comparison of Microfluidic and Turbulent Jet Mixing, Mol. Pharm. 20 (8) (2023 Aug 7) 4285–4296, https://doi.org/10.1021/acs.molpharmaceut.3c00390.

[25] M. Cárdenas, R.A. Campbell, M.Y. Arteta, M.J. Lawrence, F. Sebastiani, Review of structural design guiding the development of lipid nanoparticles for nucleic acid delivery, Curr. Opin. Colloid Interface Sci., 2023, 101705.

[26] L. Schoenmaker, et al., mRNA-lipid nanoparticle COVID-19 vaccines: Structure and stability, Int. J. Pharm. 601 (2021) 120586.

[27] I.V. Zhigaltsev, et al., Bottom-Up Design and Synthesis of Limit Size Lipid Nanoparticle Systems with Aqueous and Triglyceride Cores Using Millisecond Microfluidic Mixing, 2012/02/21, Langmuir 28 (7) (2012) 3633–3640, https://doi.org/10.1021/la204833h.

[28] D.M. Strelkova Petersen, N. Chaudhary, M.L. Arral, R.M. Weiss, K.A. Whitehead, The mixing method used to formulate lipid nanoparticles affects mRNA delivery efficacy and organ tropism, 2023/11/01/, Eur. J. Pharm. Biopharm. 192 (2023) 126–135, https://doi.org/10.1016/j.ejpb.2023.10.006.

[29] B.K. Johnson, R.K. Prud'homme, Chemical processing and micromixing in confined impinging jets, AIChE J 49 (9) (2003) 2264–2282.

[30] B.K. Johnson, R.K. Prud'homme, Flash nanoprecipitation of organic actives and block copolymers using a confined impinging jets mixer, Aust. J. Chem. 56 (10) (2003) 1021–1024.

[31] C. E. Markwalter, R. F. Pagels, B. K. Wilson, K. D. Ristroph, and R. K. Prud'homme, "Flash NanoPrecipitation for the Encapsulation of Hydrophobic and Hydrophilic Compounds in Polymeric Nanoparticles," *JoVE*, no. 143, p. e58757, 2019/01/07/ 2019, doi: https://doi.org/10.3791/58757.

[32] K.D. Ristroph, R.K. Prud'homme, Hydrophobic ion pairing: encapsulating small molecules, peptides, and proteins into nanocarriers, Nanoscale Adv. 1 (11) (2019) 4207–4237.

[33] W.S. Saad, R.K. Prud'homme, Principles of nanoparticle formation by flash nanoprecipitation, Nano Today 11 (2) (2016) 212–227.

[34] A. Haryono, M. Ghozali, S. B. Harmami, Y. Meliana, Effect of Concentration on the Ionic Interaction between Polystyrene Sulfonate and Cationic Surfactant in Aqueous Solution, in: Materials Science Forum, 2018, vol. 916: Trans Tech Publ, pp. 24–29.

[35] P. Balding, R. Borrelli, R. Volkovinsky, P.S. Russo, Physical properties of sodium poly (styrene sulfonate): comparison to incompletely sulfonated polystyrene, Macromolecules 55 (5) (2022) 1747–1762.

[36] J. Fundin, W. Brown, Polymer/surfactant interactions. Sodium poly (styrene sulfonate) and CTAB complex formation. Light scattering measurements in dilute aqueous solution, Macromolecules 27 (18) (1994) 5024–5031.

[37] V. Patel, N. Dharaiya, D. Ray, V.K. Aswal, P. Bahadur, pH controlled size/shape in CTAB micelles with solubilized polar additives: a viscometry, scattering and spectral evaluation, Colloids Surf A Physicochem Eng Asp 455 (2014) 67–75.

[38] K. Kogej, E. Theunissen, H. Reynaers, Effect of Polyion Charge Density on the Morphology of Nanostructures in Polyelectrolyte– Surfactant Complexes, Langmuir 18 (23) (2002) 8799–8805.

[39] K. Kogej, Association and structure formation in oppositely charged polyelectrolyte–surfactant mixtures, Adv. Colloid Interface Sci. 158 (1) (2010/07/ 12/ 2010,) 68–83, https://doi.org/10.1016/j.cis.2009.04.003.

[40] C.E. Markwalter, R.K. Prud'homme, Design of a small-scale multi-inlet vortex mixer for scalable nanoparticle production and application to the encapsulation of biologics by inverse flash nanoprecipitation, J. Pharm. Sci. 107 (9) (2018) 2465–2471.

[41] Y. Liu, C. Cheng, R.K. Prud'homme, R.O. Fox, Mixing in a multi-inlet vortex mixer (MIVM) for flash nano-precipitation, Chem. Eng. Sci. 63 (11) (2008) 2829–2842.

[42] R.F. Pagels, J. Edelstein, C. Tang, R.K. Prud'homme, Controlling and predicting nanoparticle formation by block copolymer directed rapid precipitations, Nano Lett. 18 (2) (2018) 1139–1144.

[43] V. Francia, R.M. Schiffelers, P.R. Cullis, D. Witzigmann, The biomolecular corona of lipid nanoparticles for gene therapy, Bioconjug. Chem. 31 (9) (2020) 2046–2059.

[44] N. Kristen, R. von Klitzing, Effect of polyelectrolyte/surfactant combinations on the stability of foam films, Soft Matter 6 (5) (2010) 849–861.

[45] S.M. D'Addio, R.K. Prud'homme, Controlling drug nanoparticle formation by rapid precipitation, Adv. Drug Deliv. Rev. 63 (6) (2011) 417–426.

[46] J.M. Goronja, A. Janosević-Ležaić, B.M. Dimitrijević, A. Malenović, D. Stanisavljev, N. Pejić, Determination of critical micelle concentration of cetyltrimethyl-ammonium bromide: Different procedures for analysis of experimental data, Hemijska Industrija 70 (4) (2016) 485–492.

[47] K. Pojják, E. Bertalanits, R. Mészáros, Effect of salt on the equilibrium and nonequilibrium features of polyelectrolyte/surfactant association, Langmuir 27 (15) (2011) 9139–9147.

[48] C. Monteux, C.E. Williams, V. Bergeron, Interfacial microgels formed by oppositely charged polyelectrolytes and surfactants. Part 2. Influence of surfactant chain length and surfactant/polymer ratio, Langmuir 20 (13) (2004) 5367–5374.

[49] G. Nizri, S. Magdassi, J. Schmidt, Y. Cohen, Y. Talmon, Microstructural characterization of micro-and nanoparticles formed by polymer– surfactant interactions, Langmuir 20 (11) (2004) 4380–4385.

[50] A. Popov, J. Zakharova, A. Wasserman, M. Motyakin, V. Kasaikin, Macromolecular and morphological evolution of poly (styrene sulfonate) complexes with tetradecyltrimethylammonium bromide, J. Phys. Chem. B 116 (40) (2012) 12332–12340.

[51] V.A. Hackley, J.D. Clogston, Measuring the hydrodynamic size of nanoparticles in aqueous media using batch-mode dynamic light scattering, Characterization of Nanoparticles Intended for Drug Delivery (2011) 35–52.

[52] M. Jaffrin, A. Shapiro, Peristaltic pumping, Annu. Rev. Fluid Mech. 3 (1) (1971) 13–37.

485b Studying Micromixing and Phase Transitions in Pharmaceutical Processes: A Phase-Field Modelling and Raman Microscopy Approach

Irene Moreno Flores, David McKechnie, Leo Lue, Javier Cardona

Introduction

Anti-solvent crystallization is a widely used purification method in the pharmaceutical industry, which exploits the difference in solubility of a solute in two miscible liquids: a solvent and an anti-solvent [1]. In this process, the local values of the supersaturation are dictated by the local compositions. Variations in the supersaturation can greatly affect the properties of the final crystalline product, and, consequently, mixing plays a major role in these processes. Unfortunately, mass transfer is not well understood in these systems, often leading to the formation of unwanted crystal phases or to undesired phenomena such as liquid-liquid phase separation (LLPS) (i.e. the separation of the solute via the formation of a secondary liquid phase).

Traditionally, mixing at the microscale has been described through Fick's law, which considers composition gradients to be the driving force for mass transfer, instead of the thermodynamically accurate chemical potential. This significant limitation prevents this model from correctly describing more nuanced aspects of mass transport, such as 'uphill diffusion' [2] — the diffusion of species against its concentration gradient — which is important in anti-solvent crystallization. More realistic descriptions can be employed, such as the less commonly used Maxwell-Stefan diffusion model, which is based on chemical potential gradients. Nonetheless, these approaches still cannot adequately describe anti-solvent crystallization as they do not consider the contribution of potential interfaces within a system to the free energy, and, thus, the species chemical potentials and mass transfer. These interfaces can appear when new phases nucleate, such as in unwanted phenomena like LLPS, which arise as non-idealities in the system lead it towards unexpected regions of the phase diagram (see Figure 1(a)).

For these reasons, these models clearly do not suffice when it comes to the modelling and description of microscale mass transfer in crystallization systems, which inherently involve multiphase phenomena. Developing a model capable of doing so is crucial for the understanding and prevention of unwanted phenomena.

Also preventing the complete understanding of anti-solvent crystallization is the lack of accurate thermodynamic data and detailed mixing measurements, which is one of the main bottlenecks for the application of more accurate models, such as phase-field models, to the pharmaceutical field. Traditionally, mixing behavior has been reported in the form of Fickian diffusion coefficients, vastly reducing a complex phenomenon to a simple, but not always representative number, and increasing the difficulty of describing the system through a simplified model.

In this work, we propose a model of nucleation and LLPS based on a combination of the coupled Cahn-Hilliard/Allen-Cahn (CHAC) phase-field model [3] with a nucleation model. Phase-field models have been extensively applied in materials science to describe precipitation and growth [4]. They base their dynamics on the free energy functions of each phase of the system, thus including in their description the interfacial free energy. This provides a much more solid thermodynamic base for modelling multiphase mass transfer, making these models robust candidates for the description of this phenomenon in the crystallization field. In this context, this work focuses on the modelling of liquid-liquid phase separation in a salicylic acid-acetonitrile binary system, studying the influence of the different model parameters on process outcomes. Particularly, the effects of the interfacial free energy coefficient, the CHAC mobilities, and the minima of the phase free energy curves will be investigated.

To support the modelling work, we have also conducted experimental mixing studies in a microfluidic device, using Raman microscopy to map diffusive mixing in ethanol-water mixtures of different compositions. We propose a new standard manner for reporting mixing data, in the form of space-resolved composition maps. These effectively constitute digital fingerprints of the mixing process, and information, such as Fickian diffusion coefficients, can be easily extracted from them.

Methodology

2D phase-field model simulations were performed, using the PRISMS library [5], coupling the Wheeler, Boettinger and McFadden (WBM) nucleation model [6] with the CHAC dynamics. Adaptive meshing and periodic boundary conditions were selected. The nucleation constants of the model were estimated through already available kinetic data for salicylic acid-acetonitrile mixtures undergoing LLPS (see Figure 1(b) (left)), following the method outlined in [7]. The remaining model parameters are estimated through optimization with available in-line images of the evolution of the system, by extracting the phase fraction and the percentiles of the particle size distribution.

The setup for the experimental micromixing studies was formed by a Harvard Apparatus PHD ULTRATM 4400 remote syringe pump, used to infuse liquid from two Hamilton Gastight 1/4-28 UNF glass syringes connected via PTFE tubing to a hydrophilic glass Y-junction chip. The channel volume and length are 0.2μL and 12.5mm, and its depth and width are 100 μm×205 μm, respectively.

Raman spectra were captured using the Horiba XploRA+ confocal Raman microscope. Linear maps were obtained by collecting spectra in 5 μm steps in the direction perpendicular to the flow every 1250 μm along the channel length, building a 2D map of the mixing process. The spectra were collected using a 532 nm laser through a 50x objective with an acquisition time of 3 s and a grating of 1200 gr/mm. They were converted into composition values through fitting with a partial least-squares method. The resulting maps, which capture the evolution of the mixing process along the microchannel, are fed to an optimization loop that allows the estimation of the Fickian diffusion coefficients.

Results and discussion

The composition and phase-variable maps obtained from the phase-field model simulations, a sample of the latter given in Figure 1(b) (right), prove the far higher suitability of this approach to describe nucleation and LLPS in the microscale within crystallization systems in comparison to more common diffusion models. The qualitative comparison of the simulated maps and the available experimental images provide sufficient reason to believe in the feasibility of this approach for the parameter estimation of the model.

The experimental microfluidic composition maps, an example of which is presented in Figure 1(c) (right), show a steady mixing of ethanol and water solutions, with the vertical composition profile flattening progressively along the channel due to the increased contact time. Additional composition maps have been captured for ethanol-water mixtures at different compositions and flow rates, as summarised in Figure 1(d). Preliminary results show an error <5% between literature and experimental values for the Fick diffusion coefficient of ethanol-water mixtures. The experiments also show the formation of an interface that disappears gradually, as seen in Figure 1(c) (left). This most likely corresponds with the laminar flow diffusion interface, commonly observed between co-flowing liquids in microfluidic systems [8]. The Raman maps allow us to monitor the composition of the interface as the liquids mix, albeit with a certain unavoidable uncertainty, especially at the edges of the microchannel, due to the complexities and nuances of the setup. Nonetheless, they successfully prove the feasibility and usefulness of the methodology, and its great potential for future applications to more complex systems.

Conclusions

In this work, we prove the suitability of describing mass transfer in binary crystallization systems through the CHAC phase-field models in combination with WBM nucleation, given the similarities between the simulated and experimental evolution of a salicylic acid-acetonitrile system. A workflow to identify the required model parameters is proposed, consisting on a combination of kinetic data and optimization via the comparison of relevant key performance indicators obtained from available experimental data. Finally, a novel experimental setup is proposed as a means of conducting information-rich mixing studies, that could be used in the future to further fit these models for multicomponent, multiphase systems. The versatility of these maps is proven by extracting Fick's diffusion coefficient for ethanol-water mixtures, with preliminary results indicating an error <5%. Overall, the proposed modelling and experimental framework has the potential to greatly enhance our understanding of mixing processes and LLPS in any chemical process involving diffusive mixing of non-ideal solutions. Ultimately, this will lead to safer, more robust manufacturing of chemical and pharmaceutical products.

References

[1] Lewis, A., Seckler, M., Kramer, H., van Rosmalen, G. (2015). Industrial Crystallization: Fundamentals and Applications. Cambridge University Press.

[2] Krishna, R. (2015). Uphill diffusion in multicomponent mixtures. *Chem. Soc. Rev.*, 44, 2812-2836.

[3] Shen, J., Yang, X. (2010). Numerical approximations of Allen-Cahn and Cahn-Hilliard equations. *Discrete Contin. Dyn. Syst, 28*(4), 1669-1691.

[4] Moelans, N., Blanpain, B., Wollants, P. (2008). An introduction to phase-field modeling of microstructure evolution. *Calphad, 32*(2), 268-294.

[5] S. DeWitt, S. Rudraraju, D. Montiel, W.B. Andrews, K. Thornton. (2020). PRISMS-PF: A general framework for phase-field modeling with a matrix-free finite element method. *npj Computational Materials* **6**, 29.

[6] Wheeler, A. A., Boettinger, W. J., McFadden, G. B. (1992). Phase-field model for isothermal phase transitions in binary alloys. *Physical Review A, 45*(10), 7424.

[7] Brandel, C., ter Horst, J. H. (2015). Measuring induction times and crystal nucleation rates. *Faraday discussions, 179*, 199-214.

[8] Lin, Y., Yu, X., Wang, Z., Tu, S. T., Wang, Z. (2012). Laminar flow diffusion interface control in a microchannel with accurate Raman measurement. *Chem. Eng. Process.: Process Intensif.*, *57*, 1-7.

(a)

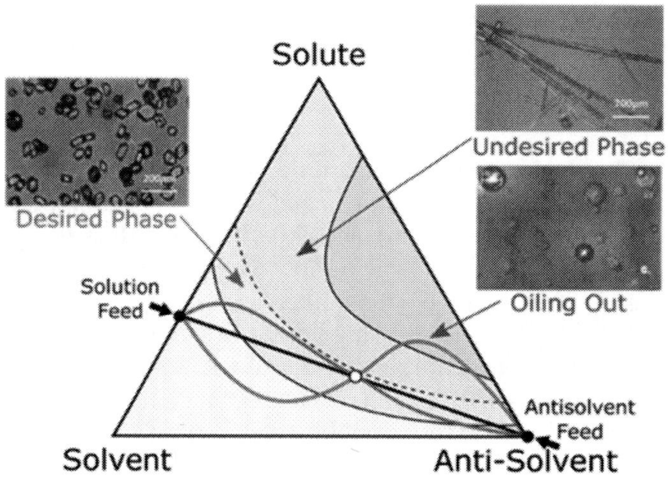

(b)

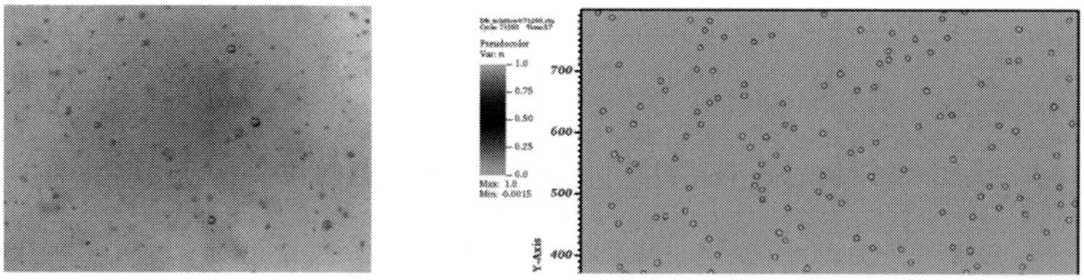

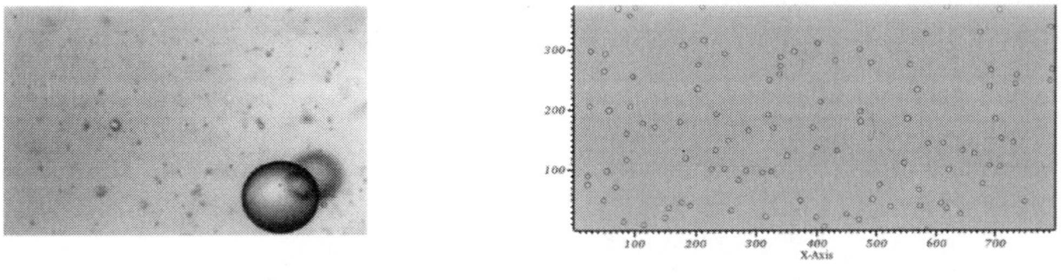

(c)

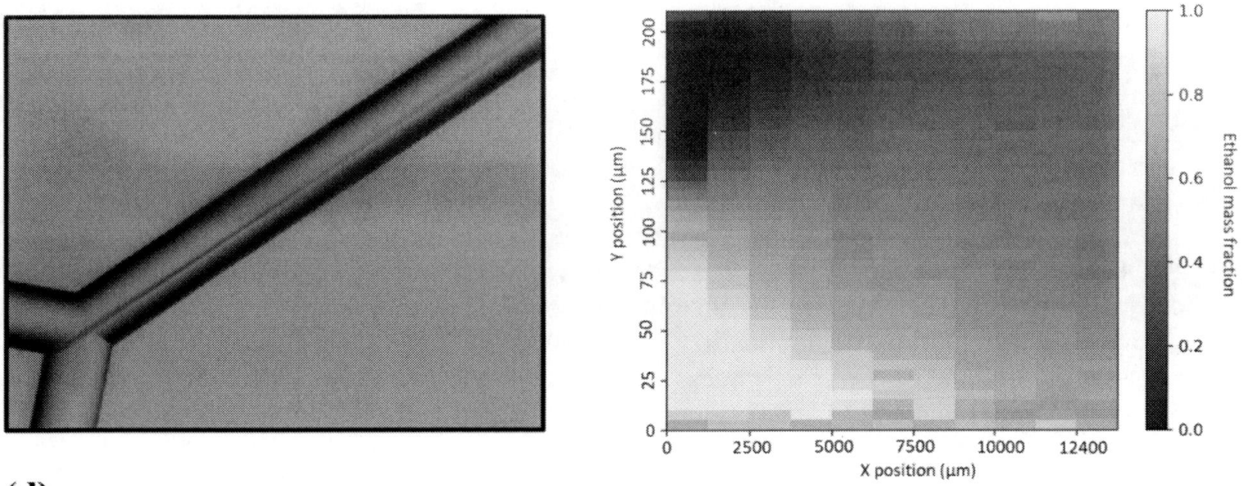

(d)

Inlet 1 (%wt EtOH)	100	75	50	25	75	50	25	100	100
Inlet 2 (%wt EtOH)	0	0	0	0	100	100	100	0	0
Flow rate (μL/min)	1.5	1.5	1.5	1.5	1.5	1.5	1.5	1	2

Figure 1. (a) Ternary phase diagram with examples of non-ideal crystallization trajectories leading to desired and unwanted regions (left) and example of the mixing map evolution during a phase transformation (right). **(b)** Experimental image of liquid-liquid phase separation in a salicylic acid-acetonitrile system (left) and snapshot of the phase fraction from the proposed phase-field model combination (right). **(c)** Optical microscope image of a 50%wt EtOH + 100% H_2O system at 5 μL/min showing the formation of an interface (left) and experimental mass fraction map obtained through Raman microscopy for a 100% EtOH + 100% H_2O system at 2 μL/min (right). **(d)** Summary of the different conditions tested in the water-ethanol mixing studies.

485c Applications of Computational Fluid Dynamics for Prediction of Buffer Dissolution in Mixing Tanks for Biopharmaceutical Processes

Behnam Partopour, Johannes Wutz, Abhinav Hazarika, John Thomas

In the quest for efficiency and precision in biopharmaceutical manufacturing, the rate and homogeneity of buffer solutions within mixing tanks emerges as a critical parameter for engineering and qualification runs and influencing the quality and consistency of the final product. This presentation delves into the innovative application of Lattice Boltzmann based computational fluid dynamics (CFD) as a predictive tool for understanding and optimizing the dissolution process of buffers in mixing tanks. Leveraging the power of MStar-CFD, we dissect the complex interplay of fluid mechanics and mass transfer to provide a detailed visualization and quantitative analysis of the mixing process, including the identification of potential dead zones and the prediction of dissolution rates under various operational conditions.

Through a series of simulations, this study reveals how variations operating conditions affect the efficiency of buffer dissolution, offering insights into optimal design and operational strategies to achieve uniform dissolution. The application of CFD allows for the a priori prediction of mixing outcomes, facilitating the move from process development to clinical and commercial manufacturing that can significantly reduce time and resources related to trial-and-error experimentation during this phase.

Moreover, this presentation highlights the role of CFD in scaling up processes from laboratory to industrial scale, ensuring that buffer mixing remains efficient and consistent across different production volumes. By integrating CFD simulations into the development and optimization of mixing processes, biopharmaceutical manufacturers can enhance product quality, reduce production costs, and accelerate the time-to-patient for vital therapeutic products. This study not only underscores the utility of CFD in advancing biopharmaceutical manufacturing but also sets the stage for its broader application in the optimization of various bioprocesses.

485d The Hydrodynamics of Vials and Laboratory Bottles

Justin Pace, Sanjaykumar Patel, Gerard Bredael, Andre Hermans, Piero Armenante

Chemistry and pharmaceutical laboratories commonly utilize vials and small bottles for variety of purposes, including the evolution of the dissolution of different mixture constituents, the dispersion of hard-to-dissolve solids in liquids, the determination or rate precipitation of crystals in reaction precipitations, and many others. Often, it is often the case that only a minimum amount of an Active Pharmaceutical Ingredient (API) is available during pharmaceutical drug development, implying that experimentation in larger scale equipment is impossible. In these small systems, mixing and agitation in general is achieved with a stir bar placed inside the bottle and stirred by a magnetic stirrer underneath. In all these small containers, the hydrodynamics plays a critical role in the process, and it is critical to have information on how the flow field varies depending on geometry and operating conditions in order to better understand mixing processes and scale scaleup. However, little information is available in the literature on the hydrodynamic and mixing performance of small bottles. Therefore, the purpose of this work was to experimentally determine the hydrodynamics of magnetically stirred vials and bottles of different sizes using Particle Image Velocimetry (PIV) and provide some guidance for scaleup. For this purpose, a custom-made vial/bottle centering equipment was built to precisely center the containers and the stir bar with respect to the stirrer. Particle Image Velocimetry (PIV) was then used to quantify all three components of the velocity vectors on a vertical plane through the bottle centerline, as well as across horizontal cross-sections, for bottles ranging in size from 4 mL to 250 mL. Different magnetic bars/bottle combinations were studied, some of which are shown in Figure 1. Additionally, the hydrodynamics in bottles with different fill volumes was also investigated. In a preliminary phase of the project different combinations of bottle size, magnetic bar size and type (e.g., with or without central pivot), and agitation speed were tested for stability of the stirring bar motion. Only the hydrodynamics of stable configurations was studied in detail. The 250-mL bottle filled with a 200-mL fill volume was selected as the reference system. This system was studied at three agitation speeds (100, 200, and 300 rpm), ranging from minimal agitation to high agitation resulting in a vortex of significant depth. Then three velocities for all the other bottles/vials were selected using the following criteria: (1) geometric similarity: experiments in bottles with nominal volumes different from that of the reference bottle were conducted at velocities that produced (approximately) similar vortex profiles as in the reference bottle (Remark: exact geometric similarity was not possible since the vial bottle/bar diameter ratio changed with bottles/vials of different sizes), and (2) constant Froude Number. Vector plots similar to those obtained in Figure 2 were obtained for all systems. The following conclusions can be drawn from this work:

- Stir bar selection is critical to ensure stability of the velocity profiles. Only certain combinations of bottles/vials and bar geometry and size result in stable motion;
- All systems studied here are extremely sensitive to minimal changes in geometry and bottle/vial/bar location;
- Even removing and reinserting the same bottle in the same system with the same agitation speed can result in small flow variations;
- Correct placement of the bottle/vial on the magnetic plate is critical;
- Careful conduction of PIV experiments resulted in reproducible velocity profiles.
- Selection of the appropriate stirring speed for any given system is important to obtain similarity of velocity profiles;
- The same Froude number should be used when scaling up/down.

Figure 1. Examples of bottles and stir bars used in this work.

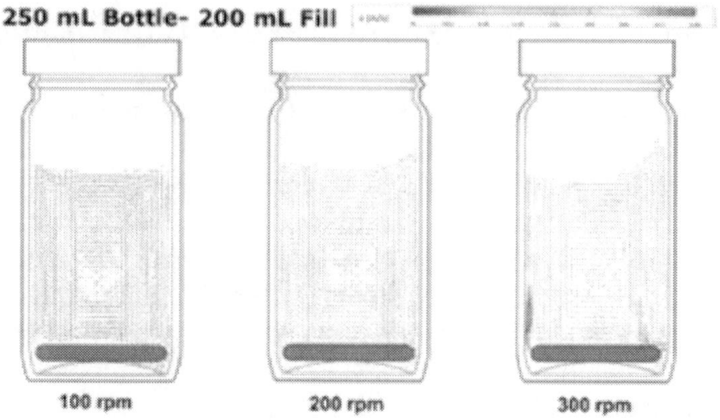

Figure 2. Velocity profiles on a vertical cross-section through the vessel centerline in a 250-mL bottled filled with 200 mL of fluid.

485e Application of a Computational Methodology for the Prediction of Mab Produced By CHO Cell Clones in pH Heterogeneous Bioreactors

Katherine Raudenbush, Jayanth Venkatarama Reddy, Luisa Attfield, Omkar Kedge, Pavlos Kotidis, Gary Finka, Giulia Marchese, Ou Yang, Sameer Talwar, Zhonghe Hu, Marianthi Ierapetritou

Monoclonal antibodies (mAbs) have become ubiquitous within the biopharmaceutical space. CHO cell-based monoclonal antibody production be challenging to scale up, and scale up optimization can begin in the early stages of cell line development. Previously, we have presented on a computational methodology for the prediction of monoclonal antibodies in large scale bioreactors, where heterogeneities in important physical and chemical parameters can be expected[1,2]. This methodology links computational fluid dynamics (CFD) modelling with kinetic models of metabolism and of a critical quality attribute called glycosylation to capture the evolution of heterogeneities while capturing their impact on cell metabolism, mAb production, and product quality.

Furthering this work, collaborators at GlaxoSmithKline (GSK) have conducted scale down simulator experiments[3] at the 250mL scale for multiple CHO cell lines, exposing them to oscillations in pH and dissolved oxygen. Data such as viable cell density, mAb titer, glucose, amino acids, lactate, and glycosylation patterns are collected in these experiments to regress the metabolic and glycosylation model parameters. By running scale down simulator experiments, we also gain information about the cell culture shifts resulting from the oscillations. Attributes which characterize the patterns of physiochemical oscillations in the experiments are incorporated into the models to account for the impact of the oscillations on metabolism and glycosylation.

Utilizing the modified kinetic models and a CFD model of a production scale bioreactor of interest, we can use the computational methodology described in previously[1,2] to iterate between CFD predictions of oscillation patterns and models of metabolism and glycosylation to predict culture dynamics, mAb production, and glycosylation patterns. Furthermore, production and product quality at the large scale can be optimized through adjusting culture conditions, specifically impeller speed, sparge composition and sparge rate, and feeding strategies.

References

1. Raudenbush, Reddy, Attfield, Kotidis, Kedge, Barodiya, Finka, Marchese, Yang, Talwar, Hu, Ierapetritou. AIChE Annual Meeting 2023.
2. Raudenbush, Thomas, Papoutsakis, Ierapetritou. AIChE Annual Meeting 2022.
3. Zakrzewski, Lee, Lye. Biotechnology Progress 2022. DOI: 10.1002/btpr.3264

485f CFD Simulation of the O_2 and CO_2 Levels in Mammalian Cell Respiration inside an Aerated Stirred Bioreactor

Harry Van den Akker, Roya Jamshidian

The cell respiration process of a Chinese Hamster Ovary (CHO) cell culture in a large-scale aerated and stirred bioreactor was modeled and simulated by using M-Star's LES software. The simulation set up was first validated with no cells present. The volume-averaged volumetric mass transfer coefficient ( ) and dissolved O_2 (DO, %) plots are in good agreement with the experimental measurement obtained using the dynamic method. Then, the cells were added to the simulation hypothetically by defining a reaction that couples the O_2 consumption rate to the CO_2 production rates. A controller was set on the pure O_2 sparger to control the DO in the tank. Some simplifying assumptions were made such that only the CO_2 variations as a result of the cell respiration were modeled. The simulation was run for almost one month on GPUs and the dissolved O_2 and CO_2 concentration evolution with time are plotted. The rates of O_2 transfer and consumption rates as well as CO_2 production and absorption rates were simulated successfully to produce results which look qualitatively correct and are in good agreement with the limited validation data available. Finally, the local dissolved CO_2 concentration data were used to calculate local and averaged partial pressure of CO_2 (pCO_2) data in the reactor which remained well in the suggested range. Overall, the simulation approach showed great potential in capturing the local heterogeneities in dissolved O_2 and CO_2 concentrations.

417a Analysis of the Benefit of Mixing for Oily Waste Sludge Biotreatment

Matthew Alexander

The biotreatment of oily wastewaters and sludges at legacy industrial refinery and petrochemical sites can be aided by the use of adequate mixing to disperse the oily phase into small separate-phase droplets, thereby enhancing interfacial area for hydrocarbon mass transfer into the aqueous phase. The increase in dispersion effectiveness directly correlates with treatment rate of the hydrocarbons, which are sparingly soluble in the aqueous phase, where bacteria effect the biodegradation. The dispersion of the oily phase material can also be aided by the use of dispersants, while mixing also commonly assists in the aeration necessary for aerobic biodegradation. This current research studies the effect that power input and mixer device type have on the dispersion effectiveness. Mixing power per unit volume of oil-containing wastewater is a critical determinant for effective oil phase dispersion. For this reason, an analysis is conducted on how effective dispersion can be achieved over a large-area hypothetical legacy industrial oily wastewater or oily sludge pond. Different approaches, such as a sizeable number of fixed mixers or aerators versus periodic movement of fewer mixers or aerators to different sections of the pond, are used to assess the power expenditure vs effectiveness of oily waste dispersion. The results of this study may be used to assist in design of biotreatment systems for oily water ponds at legacy industrial sites.

417b Characterization of Mixing and Solid Settling in Continuous Flow Slurry Reactors Using Computational Fluid Dynamics

Navraj Hanspal, Shaunak Ray, Rachel Rhodes, Mark May, Jayachandran Devaraj, Venkatramana Rajuri

Manufacturing of specialty chemicals have traditionally been performed in large-volume batch vessels with relative operating simplicity but at the cost of limited efficiency. Continuous processing offers significant advantages with respect to process intensification and scale-up but poses new challenges due to the limits in types of operations available. One major concern in batch-to-continuous process development comes from the mixing concerns related to heterogenous or multiphase reactions. The mixing considerations for batch operation differ substantially from a continuous processing approach, making the translation of technology and scale-up much more challenging. The present study investigates the mixing considerations for a slurry to slurry reaction in a scalable agitated baffle reactor (Stoli Chem, Wellesbourne, UK) which offers the benefits afforded by stirred-tanked reactors in series while still maintaining continuous production capabilities at scale. In this work we have used computational fluid dynamics (CFD) to characterize the mixing regimes and solid settling dynamics of a slurry reaction, to predict operating conditions that enable successful scale up from bench to pilot scale production of material with limited plugging concerns and comparable reaction kinetics in contrast to standard batch scale operations. Transient Large Eddy Simulation CFD models based on the Lattice Boltzmann approach were developed to characterize the flow hydrodynamics and particle suspension behavior within the flow reactor at two different length scales. Simulations were compared and validated against experimental data enabling the successful translation of reactor technology from batch operation to continuous manufacturing.

417c Improvement of Gas-Liquid Mass Transfer for Viscous Fluids in Batch Vacuum Operation

Yasuhiro Sumi, Kunihiko Matsumura, Kazuhiko Nishi

This study focuses on the removal of volatile residues from viscous fluids in a large-scale agitator reactor under vacuum conditions. The aim of this process is to accelerate chemical reactions by eliminating volatile residues or extracting such residues from polymers. The batch operation for the removal of volatile residues in a vacuum operation proceeds in three stages: a boiling phase, a transition phase, and a surface evaporation phase. During the final stage, surface evaporation occurs without the fluid boiling. Within this surface evaporation stage, the evaporation of volatile residues only occurs through the gas-liquid interface, thereby rendering the rate of mass transfer between gas and liquid crucial.

In order to augment the mass transfer at the gas-liquid interface and ensure adequate circulation of fluid within the agitator tank for sufficient renewal of fluid near the gas-liquid interface, we selected a large multi-stage impeller [1] developed by the authors. This impeller excels in vertical mixing even in medium-viscous fluids. For the purpose of promoting mass transfer at the gas-liquid interface, a large gate-type impeller has been positioned to span the gas-liquid interface. Based on insights gained from the improvement of mass transfer in the large multi-stage impeller, a large gate-type impeller has been positioned to span the gas-liquid interface on a pre-existing impeller (three-stage MIG impellers). This is advantageous for existing agitation equipment as it reduces design and equipment installation time and cuts equipment investment costs.

As a result, the installation of a large-scale multi-stage impeller equipped with an extended gate-type impeller that passes through the free surface inside the reactor, and the addition of a large gate-type impeller to the existing impeller (the MIG impellers), have led to an improvement in the rate of mass transfer at the gas-liquid interface.

The large multi-stage impeller, developed by the authors, is designed for processing polymer products with medium viscosities in the range of 0.1 to 10 Pa·s [1]. The impeller (Figure 1) comprises a rotating shaft and three vertically-oriented impellers: one two-bladed large paddle impeller and a pair of two-bladed large gate-type impellers. The shaft is vertically installed at the center of a cylindrical mixing tank. These three impellers are attached at the lower, middle, and upper stages on the rotating shaft, respectively. The paddle impeller, attached at the lower stage on the rotating shaft, is also arranged close to the bottom surface of the mixing tank. The pair of gate-type impellers are attached at the middle and upper stages on the rotating shaft, respectively. The one positioned at the middle stage is arranged at a crossing angle of 45 degrees ahead in the direction of rotation of the lower stage paddle impeller. Furthermore, it overlaps vertically with the paddle impeller. The other one, positioned at the upper stage, is arranged in the same way as the middle stage one. These three impellers have the same diameter. This impeller has the advantages of relatively low power consumption when mixing medium-viscous liquids, fewer stagnant areas within the agitator tank, and fewer vortices perpendicular to the rotation direction of the impeller, resulting in axial circulation flow.

To verify the effects of mass transfer in large gate-type impellers positioned at the top, an evaluation of the mass transfer coefficients was conducted for two types of large gate-type impeller: those submerged in liquid and those spanning the gas-liquid interface.

On the other hand, the existing three-stage MIG impellers are arranged in the lower, middle, and upper sections of the tank, respectively, each at a crossing angle of 90 degrees to the rotation direction of the impellers. Building on the knowledge gained from improving mass transfer with the large multi-stage impeller, a large gate-type impeller was installed to span the gas-liquid interface above the existing impeller.

The gas-liquid mass transfer during the surface evaporation step in a batch operation to remove volatile residues under vacuum conditions is a phenomenon where the volatile residues move from the liquid phase to the gas phase. Here, rather than evaporating the volatile substance, the volumetric mass transfer coefficient (kLa) was obtained from the pressure history of the gas phase by dissolving the volatile substance into the polymer liquid. Using the mass transfer coefficients at the gas-liquid interface for the two types of impellers - the large multi-stage impeller and the existing impeller with the additional gate type impeller - we evaluated the characteristics of mass transfer for each agitation impeller. Furthermore, we also examined the dimensionless correlation equations for each agitation impeller.

This study enabled the evaluation of mass transfer enhancement effects by various stirring impellers using kLa against the volumetric power consumption (Pv) and dimensionless correlation formulas. From the relationship between Pv and kLa in the multistage impeller equipped with the extended gate-type impeller and the multistage impeller submerged in liquid, it was found that the kLa of the multistage impeller equipped with the gate-type impeller was significantly improved. In addition, the dimensionless correlation of mass transfer was compared in the range from the laminar flow region to the turbulent flow region. Further, by separating cases by Froude number (Fr), it was evaluated using dimensionless numbers regardless of tank dimensions and fluid viscosity.

On the other hand, in the case where the gate-type impeller was added to the existing impeller, it was found that the kLa against Pv was significantly improved by placing this gate-type impeller. Furthermore, by using the relationship of Sherwood number (Sh), Reynolds number (Re), Schmidt number (Sc), and Froude number (Fr), it was also possible to estimate kLa for a commercial-scale stirred tank based on small-scale experimental data at various viscosities.

These correlations enabled the estimation of mass transfer coefficients in the final stage of volatile component removal from polymers. Based on these achievements, this method was applied in the introduction of new equipment and the improvement of existing production equipment in processes dealing with viscous fluids where mass transfer at the gas-liquid interface becomes rate-limiting.

As a result, this multi-stage impeller and the addition of the gate-type impeller have been adopted in many industrial reactors and mixing tanks in the KANEKA Group, contributing to the improvement of productivity in the functional polymers manufacturing process.

Reference

[1] Y.Sumi, M. Kaminwano: J. Chem. Eng. Japan, Vol34, No.4, 485-492(2001)

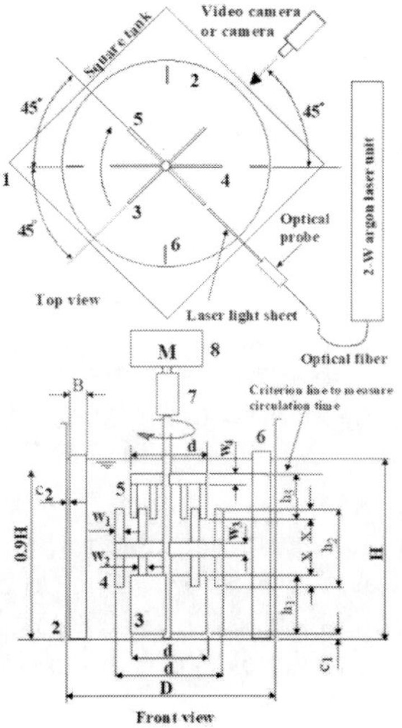

Fig. 1 Geometrical configuration of multistage impeller, four-baffles, and experimental apparatus. [1]

417d Intensification and Scale-up of Bioinspired Silica: Experimental and Computational Investigations of Mixing

Mohamed Ashar Sultan Mohamed Yousuf, Richard Hodgkinson, Carlos Brambila, Amber Keegan, Siddharth Patwardhan

Nanomaterials have a vast potential in applications and while many new nanomaterials are being discovered, their manufacturing is unsustainable and therefore their potential is not realised. With the example of bioinspired silica (BIS), we report on understanding the underpinning mixing mechanisms and develop a methodology for process scale-up and intensification. This presentation will provide experimental and computational investigations of mixing using a multi-inlet vortex mixers (MIVM) at different scales. MVIM provide a wide range of flow conditions (Re = 10-10,000) and energy dissipation ($\epsilon = 10^{-1} - 10^{3}$ W/kg), covering the lab-scales as well as plant scales, hence they are very useful in this research. The MVIM were characterised using Villermaux-Dushman reaction approaches and computation fluid dynamics. Next, the flow properties, the BIS formation process and the particle properties were investigated using a systematic experimental approach. These results (V-D and BIS synthesis) and their comparisons will be presented in order to develop mixing knowledge and scale-up rules.

145c Characterisation of Mesomixing in Laboratory Scale Reactors Using CFD and the Fourth Bourne Reaction

Niall Brady, Philip Donnellan

The role of mixing in fast, competitive-consecutive chemical reactions has been a topic of great discussion within the chemical production industry over the last thirty years[1,2,3,4,5]. A widely experienced problem in chemical reaction engineering is the production of undesired by-products through a secondary, often slow, reaction. This reduces the reaction yield of the main product and presents complication later in the products purification process. The Bourne reactions are a series of four competitive-consecutive reactions developed by Balydga and Bourne[2] to act as convenient experimental tools to study mixing efficiencies and its effect on chemical reactions. With fast competitive consecutive reactions, the composition of the reactor effluent is indicative of how well the mixing promoted the desired reactions. The Bourne reactions have been widely used to investigate the micromixing performance within a vessel, but minimal research has been carried out to assess the mesomixing performance using the Bourne reactions. Mixing can be characterised on three different scales, macro, meso and micro mixing which correspond to the bulk liquid, droplet, and molecular scale respectively. The performance at each of these scales can be predicted based on engineering correlations. These correlations are often inaccurate due to the coefficients used being unique to a specific reactor configuration. The researcher is forced to select coefficients generated in laboratory studies that most closely represent the reactor. This introduces inherent error into the calculation and can drastically reduce the accuracy of any mixing time correlation output. The meso-mixing time correlation requires a coefficient which relates the local turbulent eddy dissipation to the power per unit mass.

The Bourne reactions have been used in many experimental investigations since their conception and the first three reactions are well detailed by Balydga and Bourne[2] and Paul et al[3]. This study focused on the fourth Bourne reaction. This reaction is a competitive hydrolysis and neutralisation reaction where the fast reaction consists of HCL reacting with NaOH and the acid-catalysed decomposition of 2-2-dimethoxypropane as the slow reaction. Experiments were carried out to investigate the effect of addition location, addition rate, agitation rate, impeller type and baffles on the production of methanol through the decomposition of DMP. The concentration of methanol was used as an indicator of mesomixing performance, and these results were compared with the mesomixing timescales predicted by CFD simulations. FT-IR and GC were used to quantify the concentration of methanol in each experiment. The validity of using FT-IR to provide in-process estimations of product quantification was proposed and compared with the results of offline GC analysis.

The local turbulent eddy dissipation and local velocity values within 1 L laboratory scale OptiMax reactors were estimated using computation fluid dynamics (CFD) and used to predict the mesomixing performance using the turbulent dispersion theory. Correlations were then developed to allow these local values to be estimated without the need for CFD simulations. These correlations were compared with those currently used in small molecule laboratories and considerable variance was observed between the results. The local values are heavily dependent on addition location, agitation rate, impeller type and scale. Twenty experiments involving the competitive-consecutive Fourth Bourne reaction mechanism were carried out to investigate the relationship between mesomixing time, predicted using CFD simulations, and the yield of methanol from the secondary acid hydrolysis reaction. The yield of methanol was quantified using gas chromatography (GC) and monitored in-process using Fourier transform infrared spectroscopy (FT-IR). A linear relationship was determined between the results of GC analysis and FT-IR indicating that FT-IR could provide an accurate, in-process method of product yield quantification.

Well fitted models, that describe that design space of this study, were developed using JMP® statistical software for the coefficients α and β. These coefficients were used to describe the relationship of the average TED and impeller tip speed to the local values of TED and velocity, respectively. Currently, there exists limited literature to describe these coefficients in small lab-scale reactors. In small molecule laboratories, generalised values for 18 L are used to provide estimates for these coefficients[6]. This study determined that these generalised values are widely inaccurate in their estimation of local values in 1 L OptiMax reactors. Values for ε_{local} in OptiMax reactors were found to be greatly overestimated by the generalised values obtained by Verschuren et al[6]. The prediction expressions obtained from the JMP® models were developed into Excel tools for use in small molecule laboratories to determine more accurate estimations of local turbulent eddy dissipation and local velocity values in OptiMax reactors.

Mesomixing constants, Λ_{meso} and t_{meso}, were calculated for each simulation using engineering equations described by the inertial convective mechanism. For each simulation, addition along the surface plane resulted in a longer Λ_{meso}. This is important for rate sensitive reactions such as the Fourth Bourne reaction which was investigated in this study. A longer mesomixing length scale resulted in a larger localised zone of high HCL concentration during the experimental process (Figure 1). The values for t_{meso} increased proportionally to the values of Λ_{meso}, and a combination of fast addition and slow agitation rates resulted in high values for mesomixing time. Simulations which experienced complex transitional flow (50 rpm) produced inconsistent results and it was hypothesised that the inertial convective mechanism is valid only for turbulent flow conditions. More research is required to accurately predict the mesomixing time and length scales under transitional flow regimes using engineering correlations.

A DoE was employed to identify significant mixing parameters using the quantification of methanol from the Fourth Bourne reaction. Off-line gas chromatography analysis was used to quantify the concentration of methanol produced by the secondary acid hydrolysis of DMP reaction and thus the efficiency of mixing was quantified. A standard least squares screening model was built in JMP® from the predetermined DoE. It was determined that agitation rate and the second order interaction between agitation rate and addition rate were the most significant factors to the production of methanol in the reaction set-up. An increase in agitation rate resulted in a decrease of methanol concentration and the lowest methanol concentration observed was a result of high agitation rate and high addition rate. The plane at which the HCl was added was deemed to be of significance to the amount of methanol produced. Addition close to the impeller reduced the mesomixing length scale and resulted in lower concentrations of methanol compared to addition along the surface of the liquid. These experimentally deduced relationships have been summarised and presented in flowchart form, as seen in Figure 2, to enable the user to control the products of a competitive-consecutive reaction.

The validity of FT-IR in providing accurate, in-line estimations of product concentrations was investigated. A well fitted linear correlation with an RSq of 0.8 was observed when the peak area change in IR was compared with the methanol concentration determined using off-line GC analysis. Strong experimental evidence was observed to suggest that FT-IR could be used to estimate product concentration through the creation of a "standard curve" between peak area change and product concentration for a known product.

The results of this study have highlighted several areas that have the potential for further research. It was discovered that use of the inertial-convective mechanism to predict the mesomixing time and length scale misrepresented transitional flow regimes, and some scenarios of addition to low velocity regions of flow. These scenarios resulted in outliers when the mesomixing time predictions were compared with the concentration of methanol produced. The removal of these outlier improved the linear relationship more than two-fold and this accentuates the potential for further research into the determination of mesomixing constants under transitional flow regimes. Further work can also be carried out to improve the RSq value between the mesomixing time CFD predictions and the product concentration in competitive consecutive reactions under turbulent flow conditions.

The validity of using FT-IR to provide in-line product concentration estimates is an area of huge potential significance in the control of competitive reaction systems in small molecule laboratories. FT-IR is generally considered a crude tool used to provide an indication of changes within the IR spectrum over the duration of the process. The results of this study showcase potential for the use of FT-IR to provide accurate, in-line product yield data. Through the development of a "standard curve" using peak area change from FT-IR and concentration data determined using GC, the concentration of a known product can be interpolated to provide an accurate indication of the reaction status. This could be of great significance to the optimisation of product quality in fast, mixing-sensitive reactions such as competitive-consecutive reactions, precipitation reactions and fermentation processes.

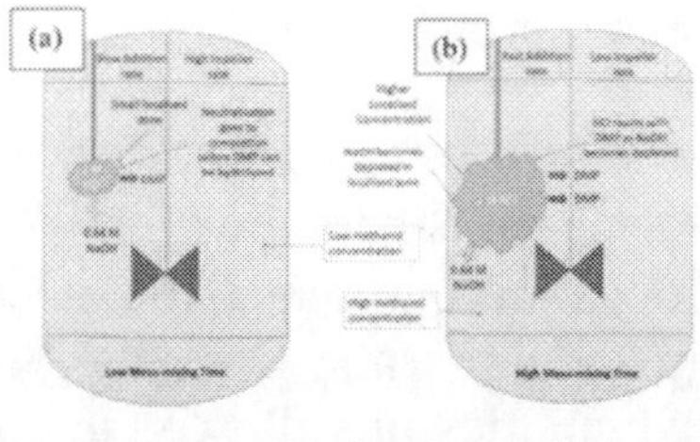

Figure 1: *Visualisation of the impact of (a) slow addition and high agitation rates and (b) high addition and low agitation rates on the mesomixing constants and products of the Fourth Bourne reaction*

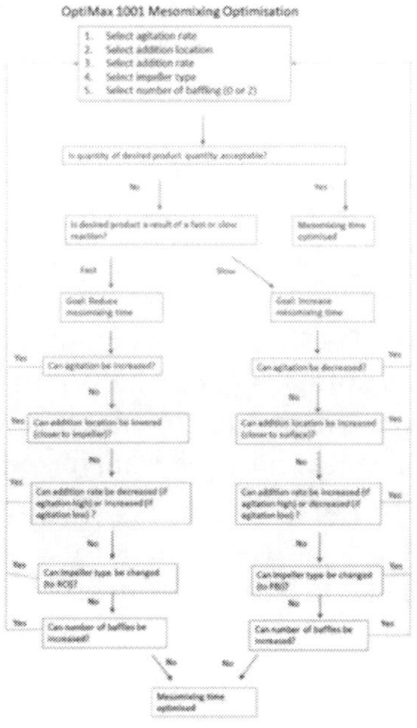

Figure 2: *Workflow to control products of a competitive-consecutive reaction*

417g G-Value: Is it genuinely useful in sizing Flocculators and Rapid Mixers?

Richard Grenville, Jason J. Giacomelli

G-value is a shear rate which is used for the sizing of mixing equipment for flocculant dispersion and floc growth for water treatment. It has several advantages in that it is easily calculated and provides a gross representation of mixing in the vessel which is fairly well correlated with performance. But it also has a major disadvantage in that it does not take spatial variations of shear rate into account. The local G-value measured at the tip of the impeller blade will be higher than that measured near the vessel wall. Research has shown that it is the local G-value at the impeller tip which determines the floc size that is generated by the mixing equipment.

There are two steps in the flocculation process. In the first a flocculant is added which is mixed with the suspended solids. This is the Rapid Mixing step, and the flocculant eliminates the electrostatic forces between the particles which prevent them from colliding and adhering. Once the flocculant is mixed, the particles must be transferred to an environment in which they are allowed to collide and for the flocs to grow. This is the Flocculation step and there must be sufficient fluid motion to promote particle-particle collisions but not so much that the growing flocs are broken. Ultimately the balance between flow and shear determines how successful the flocculation process will be.

The common method for defining the intensity of mixing by specifying a G-value first proposed by Stein & Camp (J. Bos, Soc. Civ. Engrs., 1943). G is dependent on the power input per mass of fluid by the agitator's impeller and the kinematic viscosity of the fluid. It can also be expressed in terms of the power input by the impeller and the volume and dynamic viscosity of fluid. Its value is specified by the consulting engineer. A high value of G is chosen for Rapid Mixing, typically 500 - 1000 s^{-1}, while a low value is chosen for Flocculation, typically 50 - 100 s^{-1} (Waste Water Engineering, Fifth Edition, McGraw-Hill Education, 2014).

Impellers in agitated vessels can be characterized in terms of their ability to generate flow and shear. This understanding can be applied to determine which geometry is most efficient for rapid dispersion of flocculant and which is most efficient for promotion of floc growth. The shear generated by a small diameter hydrofoil will be higher than the shear generated by, for example, a large diameter pitched-blade turbine at the same G-value and power input. But a small hydrofoil will operate at a higher speed to deliver the required power input, based on the specified G-value. It is attractive to equipment vendors to quote the higher speed agitator because the shaft torque will be smaller making the equipment size and cost lower and their bid more attractive commercially, but this will not necessarily be the best option for the process. If the shear is too high the operator can compensate by using more flocculant adding to their running costs.

568a North American Mixing Forum Award Lecture - All Mixed up: An Academic's Three-Decade Adventure in Industrial Mixing Practice

Kevin Myers

All Mixed Up: An Academic's Three-Decade Adventure in Industrial Mixing Practice

How did an academic manage to survive in industry for thirty years? This presentation will describe how a successful collaboration between academia and industry started and was sustained over the long haul. Reasons for establishing the relationship and keys to making it work, including the role of the larger mixing community, will be reviewed. Highlights of the work performed, including the people involved, will be revisited and efforts to address some puzzling results will be discussed.

AUTHOR INDEX

AIChE
120 Wall Street, FL 23
New York, NY 10005-4020

ISBN 979-8-3313-1664-8

9 798331 316648

CURRAN ASSOCIATES INC.
proceedings.com